D0426205

THE END OF ANIMAL FARMING

THE END OF ANIMAL FARMING

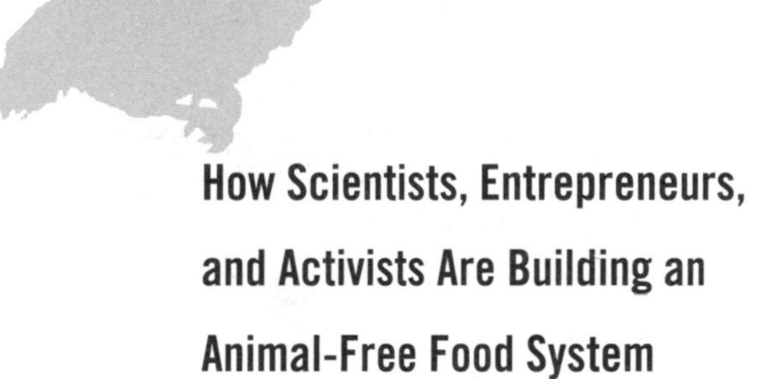

How Scientists, Entrepreneurs, and Activists Are Building an Animal-Free Food System

JACY REESE

Beacon Press, Boston

BEACON PRESS
Boston, Massachusetts
www.beacon.org

Beacon Press books
are published under the auspices of
the Unitarian Universalist Association of Congregations.

Printed in the United States of America

21 20 19 18 8 7 6 5 4 3 2 1

This book is printed on acid-free paper that
meets the uncoated paper ANSI/NISO
specifications for permanence as revised in 1992.

Text design and composition by Michael Starkman
at Wilsted & Taylor Publishing Services

Library of Congress Cataloging-in-Publication Data

Names: Reese, Jacy, author.
Title: The end of animal farming : how scientists, entrepreneurs, and activists are building an animal-free food system / Jacy Reese.
Description: Boston : Beacon Press, [2018] | Includes bibliographical references and index.
Identifiers: LCCN 2018009613 (print) | LCCN 2018011269 (ebook) | ISBN 9780807019474 (ebook) | ISBN 9780807019450 (hardcover : alk. paper)
Subjects: LCSH: Food animals—Moral and ethical aspects. | Food supply—Moral and ethical aspects. | Meat substitutes—Economic aspects.
Classification: LCC HV4757 (ebook) | LCC HV4757 .R44 2018 (print) | DDC 179/.3—dc23
LC record available at https://lccn.loc.gov/2018009613

This book is dedicated to ancient changemakers
like Pythagoras (vegetarian and Greek philosopher)
and Ashoka (animal advocate and Indian emperor),
whose ripple effects plausibly did more good
than anyone else's in
the history of the world.

CONTENTS

Introduction ix

1. The Expanding Moral Circle 1
2. Emptying the Cages 17
3. The Rise of Vegan Tech 38
4. How Plant-Based Will Take Over 57
5. The World's First Cultured Hamburger 73
6. The Psychology of Animal-Free Food 95
7. Evidence-Based Social Change 113
8. Broadening Horizons 129
9. The Expanding Moral Circle, Revisited 145

Acknowledgments 165

Notes 167

Bibliography 193

Index 203

INTRODUCTION

This is not a book about the problems of animal farming. Scores of compelling books, documentaries, news articles, and scientific papers have detailed the damage animal farming does to public health, agricultural workers, rural communities, the national economy, the global food supply, our air and water, and, of course, farmed animals.[1]

This is a book about exactly how we can solve those problems.

So much has been written exposing and condemning the animal agriculture industry that technology theorist Tom Chatfield listed "eating meat and factory farming" first in his predictions of what our descendants in centuries to come will deplore about today's society. Journalist Ezra Klein, author Steven Pinker, business magnate Richard Branson, science educator Bill Nye, and Indian politician Maneka Gandhi have all made similar forecasts, based primarily on concern for farmed animals.[2] In 2017, the BBC even produced the mockumentary *Carnage* based in the vegan world of 2067. It took a critical and humorous look at Britain's unpleasant history of eating animals.

It might seem surprising that the plight of these neglected creatures—by numbers, around 93 percent of farmed animals are chickens and fish[3]—is so compelling an issue, given the fact that humans are still plagued by disease, oppression, war, racial and economic inequality, and other pressing social issues. However, even if we ignore the harms animal farming causes to humans, a bit of reflection exposes a compelling moral urgency. Consider these three facts: First, there are over one hundred billion farmed animals alive at this moment—more than ten times the number of humans.[4] Second, over 90 percent (over 99 percent in the US) of these animals live on industrial, large-scale "factory farms" enduring atrocious cruelty such as intense confinement in tiny cages, brutal mutilation and slaughter methods, and rampant disease and suffering from artificial breeding for excessive production of meat, dairy, and eggs.[5] Third, today we have scientific consensus that these are sentient beings with the capacity to feel great joy and suffering.[6] If we put these facts together, then we see animal farming as more than an abstract system of machinery and livestock. Animal farming is the moral catastrophe of one sentient being with a heartbreaking life story, plus another sentient being . . . plus another . . . plus another . . . plus another . . . more than one hundred billion times. Historian Yuval Noah Harari, author of the books *Sapiens: A Brief History of Humankind* and *Homo Deus: A Brief History of Tomorrow*, went so far as to suggest that animal farming is "the worst crime in history."[7]

You might feel cautious about this opposition to all animal farming: If there's a small percentage of the animal agriculture industry that's not factory farming, shouldn't we support that segment instead of abolishing the whole thing? What's wrong with buying eggs from happy hens? I will share my views on this topic at length in chapter 6, arguing that we should oppose all animal farming, but note that most of this book's arguments don't rely on that viewpoint. Food advocates of all viewpoints are still united against the vast majority of modern animal farming, and you could read the rest of this book as *The End of Factory Farming* with little issue.

The future is brighter than you think

Some people dismiss the notion that animal farming could ever end, arguing that meat, eggs, and dairy are too deeply embedded in human society and human nature. To the contrary, humanity has already seen many similar transitions. With animals, we shifted from whale oil to petroleum to light our homes. We shifted from horse-drawn carriages to automobiles as the dominant mode of local transportation. We're currently seeing a major transition in circuses from animal-based to human-based entertainment. Outside of animal issues, we've seen dramatic changes in our relationships with other humans, many of whom have been treated with blatant insensitivity similar to what animals currently endure. The inhumane practices of child labor have been largely abolished. Women, people of color, the mentally disabled, and many other oppressed groups have seen radical shifts from being viewed as mere property or lesser citizens to being appreciated as individuals with their own rights, desires, and interests—though of course society still has a long way to go.

In the specific domain of animal farming and vegetarianism, we've already seen a remarkable transition that suggests that species differences, like racial and gender divisions between humans, are not unbreakable barriers to moral consideration. What's more, our dietary habits are not set in stone or "hardwired" into humanity: they are more determined by cultural mores. In the West, vegetarianism was for centuries seen as a strange, puritanical lifestyle. In the 1800s, Sylvester Graham and John Harvey Kellogg—a minister and doctor now best known for their crackers and breakfast cereal, respectively—advocated a bland vegetarian diet to improve health and reduce sexual desire and immoral tendencies. (Other recommendations included vibration therapy, complete abstinence from alcohol and tobacco, and frequent enemas.)[8] While this made vegetarianism popular with the most ascetic and health-crazed segment of the population, it locked vegetarianism into the US public consciousness as a fringe diet for the self-righteous and holier than thou.

Modern vegetarian and vegan diets retain remnants of this stereotype, but they are taken far more seriously today because they are adopted not for personal purity but for the sake of animals, the environment, and human welfare. When a startup called Memphis Meats released the world's first cultured meatball in 2016—real meat made with animal cells grown in cell culture (details in chapter 5), the company announced plans to launch its products at barbecue restaurants in the Memphis area, appealing directly to the most carnivorous demographics.[9] World-class athletes like mixed martial artist Nate Diaz, strongman Patrik Baboumian, tennis players Serena and Venus Williams, and NFL defensive end David Carter are all vegan, and they credit their vegan diets with enhancing athletic performance.[10]

Animal-free foods are increasingly associated with exciting new technology, Silicon Valley, and intellectual trendsetters. Consider the role of a single company, Google. Google's cofounder, Sergey Brin, funded the creation of the world's first cell-cultured hamburger in 2013.[11] In 2015, Google tried to buy one of the leading plant-based food companies, Impossible Foods, though Impossible wasn't willing to sell.[12] Eric Schmidt, the chairman of Alphabet, Google's parent company, listed "nerds over cattle" as the most important tech trend in 2016. He said animal-free foods would take over the conventional meat industry in the coming decades.[13] Even internally, Google has been an early adopter of this trend in its employee cafeterias, replacing animal-based products with animal-free foods such as Just Mayo, a popular brand of eggless mayonnaise, and New Wave Foods' plant-based shrimp, which is made from the same algae that marine shrimp eat.[14]

The ace in the hole for the inevitability of the end of animal farming is the incredible inefficiency of making meat, dairy, and eggs from animals. Farmed animals consume calories and nutrients from plants, and they use that energy to do a lot more than produce meat, dairy, and eggs.[15] They have all the normal bodily functions like breathing, movement, and growing by-products like hoofs, organs, and hair. These processes mean farmed animals have a caloric conversion ratio of 10:1 or more. For every ten calories of food we feed them, we get only about one calorie of meat in return. And for every ten grams of plant-based protein, we get at most two grams of animal-based protein.[16]

Culinary professionals and food scientists are increasingly cutting this waste by taking the constituents of animal products (fats, proteins, nutrients, water) directly from plants and assembling them into the architecture of meat. They can also make cultured meat like the Memphis Meats meatball, real animal flesh made by using cell cultures to grow meat in the same process that happens inside an animal's body, so it's molecularly identical to conventional meat. The efficiency of both these processes suggests that, in the long run, humanity will use its technological prowess to make meat, dairy, and eggs directly without the wasteful process of raising and killing animals, regardless of ethics. To put it another way, a big reason we're going to see the end of animal farming is that it doesn't have to mean the end of meat.

Effective altruism

My personal perspective in writing this book begins with growing up in rural Texas. The first time I saw a goat give birth is imprinted in my memory. It was a bloody ordeal, and all the neighbors circled around. They gathered to ensure that the mother was safe, to ensure the kid was healthy, and frankly because there just wasn't that much else to get excited about in the woods of rural East Texas. In middle school, I made a "goat exerciser" for the science fair. My classmate and I assembled PVC pipes into a merry-go-round and leashed a goat to each side so they could walk in circles together. At other times in my childhood, I helped raise chickens and rabbits, competed in the county fair, and grew my own patch of watermelons and okra in a forest clearing behind the house my dad built himself in true Texas fashion.

Like many American kids, I was also raised with coloring books that showed happy, healthy farmed cows, pigs, and chickens on green grass under a blue sky. I believed that animal agriculture was a decentralized industry composed of thousands of Old MacDonald farms dotting the landscapes of rural America. Every day on my way to school, I passed by dozens of cows lounging on lush pastures chewing grass and cud. To me, meat was an embodiment of the natural, pristine connection between animals and humans who lived off the land.

One day when I was fourteen, I found a website describing the cruelties endured on the standard factory farm, such as that the average chicken is afforded less space than a sheet of printer paper. I found out that most of the meat I ate was from these farms, not the farms I saw on the roadside. So I put two and two together and went vegetarian on the spot.

My reasoning for that decision was based on taking action that results in the greatest positive impact, a philosophy now known as "effective altruism." We've seen a social movement emerge based on this philosophy where the community applies evidence and reason to benefit others to the greatest extent possible. In fact, every research question in this book is motivated by the ultimate question of how we can do the most good.

In high school and college, my interests were centered on the other social issues effective altruists prioritize, like fighting malaria and ensuring the safety of dangerous technologies like nuclear weapons and artificial intelligence. However, a central tenet of the effective altruism mind-set is to stay open-minded and keep an eye out for the so-called Cause X, a new way of doing good that could be even better than your current strategy. Eventually I became convinced that reforming the food system was my Cause X, based on three criteria: its scale, tractability, and neglectedness.[17]

Since then, I've committed my career to applying the lens of effective altruism to the issue of animal farming. I'm a participant-observer who sees my role in social movements as supplying other advocates with the tools they need to succeed. I see this book as a road map for the farmed animal movement to improve the food system as quickly and reliably as possible. This book outlines the successful strategies already being implemented by movement forerunners and sketches a path forward based on careful analysis of their work as well as evidence from fields such as history, psychology, sociology, and marketing. The most basic argument of this book is that we can take the scientific approach used in fields like medicine and apply it to social change. We can diagnose a social movement, assess its strengths and weaknesses, and prescribe evidence-based solutions.

If you are already a part of the movement, I hope this book will show you a path toward the brighter future you desire. If you have yet to join, I hope this book will inspire you to take up arms, whether in your everyday life, as a thoughtful consumer and citizen, or in the work you do, as an activist, entrepreneur, or scientist. In either case, I hope that the evidence in this book will change your mind in some important ways, just as writing it has changed mine. One of the most useful skills advocates can develop is a sincere satisfaction in changing their mind, putting the goal of effectiveness before the goal of having been correct.

As humanity's power grows, it will become increasingly important that humanity considers the interests of all sentient beings, not just those who can vote or organize on their own behalf, and that we approach complex social issues like animal farming with the perspective of effective altruism. The moral arc of the universe bends toward justice, but only if people like us take hold.

Sketching out the road map

In the first two chapters of this book, I will outline a history of humanity's expanding moral circle and the current state of the farmed animal movement. These chapters address how humans have come to think of animals as sentient beings with their own desires and interests; how this has led to an unstable dissonance between our values and the state of our food system; which advocacy strategies have been most successful so far, such as the exposure of rampant animal cruelty through undercover investigations; and two popular advocacy strategies that I believe have been woefully counterproductive.

The third and fourth chapters examine the history and limited success of currently available animal-free foods, such as the classic "tofurkey" holiday dish and the recent so-called "bleeding" veggie burgers that fool blind taste-testers. These chapters tell the complicated stories of industry leaders who have one foot in activism and the other in running a competitive business. I will tackle strategic issues like the risk of corporate interests co-opting the farmed animal movement, which sort of food companies can make the biggest difference in our food system, and how we should label these foods: should we call it meat if there's no animal slaughter?

The fifth chapter is a deep dive into the new cellular agriculture industry where meat, dairy, and eggs are grown with cell cultures instead of produced via the raising and killing of animals. This industry promises to deliver not just plant-based versions that fool blind taste-testers, but actual animal products that are molecularly identical to animal flesh. I will introduce the scientists and business leaders behind this new technology, examining which strategies they can utilize to maximize its impact.

An animal-free food system depends not on technology alone, but also on changing hearts and minds. Chapters 6, 7, and 8 lay out the key strategies of activism and social change that will beckon the animal-free food system. Chapter 6 looks at why people eat animal-based foods and proposes strategies that overcome each justification, especially the notion that animal-based foods are natural. This chapter also addresses the issue of so-called "humane" animal farming, arguing that the movement needs to oppose even the animal farms that, when considered individually, seem morally acceptable. Chapter 7 tackles changing our food system on a society-wide scale, arguing for institutional over individual change, focusing on the *how* of animal-free food over the *why*, putting stories before statistics, and maintaining a cautious approach to the use of confrontation. Chapter 8 zooms in on one crucial big-picture issue: given the way vegetarian and vegan diets are currently perceived as the personal choice of white, wealthy, Western liberals, how can we transition into a global, cross-cultural mass movement?

Finally, chapter 9 zooms out to the broader context of humanity's moral circle, speculating on what the end of animal farming will look like, when it will happen, what it will mean for human-animal relationships, and what struggles sentient beings will face after the end of animal farming. I'll discuss the psychological tension advocates face between everyday struggles and the society-wide, centuries-long fight to expand the moral circle. Advocates need to keep an eye to the far future, positioning their ideological descendants to have the best shot at finally achieving a safe, just future for all sentient beings.

1. THE EXPANDING MORAL CIRCLE

IN 2015 some friends and I visited a small, picturesque farmhouse in California's San Joaquin Valley, surrounded by never-ending vineyards stretched under an open sky. We drove down a rocky, winding road to the metal gate at the front of the property. Our host, Christine, greeted us in a golf cart packed with rakes, trash bins, and other maintenance supplies. Christine is a cheerful, down-to-earth woman with a big smile who earned her degree in poultry science from Texas A&M, an agriculture- and engineering-focused university an hour away from my hometown in rural Texas.

But Christine isn't a farmer or rancher. She manages Harvest Home Animal Sanctuary, a nonprofit organization that rescues and cares for abused animals: mostly chickens, turkeys, goats, ducks, and pigs. Farm sanctuaries like hers provide lifelong care for a few fortunate animals, educate the public, and build the advocacy community through volunteer opportunities that give people a direct connection to the animals these organizations are trying to protect.

At the gate, Christine welcomed us with a beaming smile and suggested we park in the dusty clearing in front of the house centered on the property. To the left of the driveway was a dark, hay-filled pig barn where several porcine residents wagged their tails next to the fence, eager to sniff and oink at the new visitors. To my right was a small, open shelter in the middle of a large, grassy field with goats, chickens, and a couple of outspoken turkeys.

The first resident I met (who ambushes every visitor) was Boyfriend Ben, a broad-breasted bronze turkey. Ben shuffled back and forth in front of me with his feathers extended outwards in a *Look at how big I am* mating dance. He single-mindedly shuffled next to me for my entire visit, even trying to stay in my field of vision when I turned away. I tried to understand this behavior by comparing it to a man constantly flexing his muscles, but I'm not sure if I can fit the turkey-shaped peg into a human-shaped hole.

In the backyard of the farmhouse, I met Ray, a special-needs goose. Ray is blind and suffers from angel wings, in which the last joint of the wing is fixed outwards, a condition common in domestic fowl, caused by poor diet such as the overconsumption of bread. This takes away the bird's ability to fly and even makes walking difficult. I imagine it must be like having to carry a sideways two-by-four that keeps bumping into things and throwing you off balance, though surely this human-animal comparison is also imperfect. Ray has an insatiable curiosity that makes the pejorative "bird-brained" less applicable than "brainy bird." Ray especially loves splashing around in her kiddie pool and draping her neck around friendly humans.

After a few more introductions to the locals, I got to work. Harvest Home hosts monthly tours where you can lounge about and relax all day, but it's better to go there as a volunteer. For animal advocates hoping to recharge, the best experience comes from taking an active role in the animals' healing process—seeing the stark differences in health and behavior between newcomers and longtime residents. Doing so gives you a deeper appreciation of who they are and what they've been through.

At the last barn I cleaned, I met Hayden, a rowdy young pig who's always first to the fence when visitors arrive. He's actually a Yucatan minipig, raised to mature quickly and stay much smaller than wild or farmed pigs. Yucatans are favored in biomedical research for these features and because they are so docile and trusting—much like beagles, the most common dog breed used for research. Hayden was used for skin testing because of his hairlessness and his skin's biological similarity to that of humans. I don't know exactly what kind of skin research he was used for, but to be honest, I don't need to know. The burn marks and scars covering his body tell me enough.

While cleaning the pig barn, I carried used hay away in trash bins. Once while I was carrying two bins this way, Hayden trotted along behind me. I set one bin down so I could empty the other onto the compost heap. While I had my back turned, Hayden pushed his body into the bin I set down, causing it to topple over. I appreciated the assistance, but he wasn't doing it for me. For the rest of the afternoon, he dug in the hay from that bin, snorting and wagging his tail vigorously. I'm still not sure what he was looking for—perhaps an egg left over from cleaning the chicken barn, or perhaps the satisfaction of emptying a bin on his own—but I felt immensely grateful to Christine that this troublemaker had the opportunity to do whatever it was he wanted to do.

As ambassadors for their species, animals like Hayden give you a visceral appreciation of their capacity to have meaningful experiences of pleasure and suffering. Combine this understanding with the knowledge of the cruelty these animals endure in farms and slaughterhouses—as well as secondary concerns of sustainability, antibiotic resistance, the treatment of slaughterhouse workers, and numerous other issues—and it's easy to understand why an entire social movement has emerged working toward an animal-free food system.[1] But to understand the momentum and timing of the modern movement, we must first understand the historical trajectory of humanity's concern for animals.

Animal machines

In the seventeenth century, renowned philosopher René Descartes wrote that all nonhuman animals, including dogs, cats, and pigs, completely lack the ability to feel pleasure and suffering. Some historians believe Descartes held a more nuanced view—that animals have the mechanical ability to feel, but they lack *true* consciousness of those feelings—but in either case, the father of modern Western philosophy brutally illustrated his perspective when he nailed living dogs to wooden boards and hacked them to pieces. He used this experiment as evidence for his cruel philosophy, having found no evidence of a soul inside the poor dogs' bodies.[2]

The proposition of animal consciousness, although treated as common sense by many of us today, is an inherently challenging concept. It threatens our human need to feel unique in the world. The idea that humans are on top of the ladder of nature dates back to the Greek philosophers Plato and Aristotle, who believed in a natural, linear order to the world. Humans sit comfortably above all animals and plants, surpassed only by angels and other heavenly beings.[3] If Greek philosophers rejected this model and accepted that animals matter like we do, that would spark a number of pressing, uncomfortable questions. If animals are just complex machines like printers or cell phones, does that mean we are too? When we see animals die, does that prove our own mortality? And if their lives consist of simple sustenance and reproduction, can we hope for something more?

These existential difficulties contributed to the mistreatment of animals in more dire ways than the rhetoric of philosophers. In ancient Rome, humans frequently fought animals in combat, either as a form of capital punishment or as entertainment. For the opening of the Colosseum in AD 80, up to nine thousand animals, from hares to elephants, were killed in a one-hundred-day celebration.[4]

Other historical practices seem even more sadistic and absurd, such as bearbaiting (tying a bear to a post and commanding a number of dogs to attack him) and cat burning (hoisting live cats onto a bonfire in a public spectacle), both of which have been popular forms of entertainment in human history.[5]

These public acts of animal cruelty would be absurd, even terrifying, in much of modern society. Modern concern for animals is widespread. In 2015, animal rights had the highest support of any cause on the popular petition website Change.org.[6] A Gallup poll in the same year indicated that 32 percent of Americans believed "animals deserve the exact same rights as people to be free from harm and exploitation." Another 62 percent believed they deserve "some protection." Only 3 percent thought they "don't need much protection."[7] As of 2014, all fifty US states have felony provisions for animal cruelty.[8]

This is a far cry from Descartes's view that animals deserve zero protection.

The change in attitudes with respect to animals is of the utmost importance for understanding the future of food. Most animal products today come from factory farms: industrial facilities brimming with blatant abuse. Many pigs are kept in crates so small they can't turn around.[9] Fish often die by being crushed or suffocated out of water.[10] Hens lay supersized eggs almost daily, leading to a host of health issues and chronic pain, and they live in tight cages for virtually their entire lives.[11]

Some argue that these practices are just the way things are. They say that society has grown to recognize human equality across gender, race, and nationality, but nonhuman animals are too distant and different to receive our full moral consideration. We see it in our everyday language when we use "it" to refer to animals in the same way we'd refer to an inanimate object, when we use expressions like "More than one way to skin a cat" without a second thought.

I understand this skepticism, but in the rest of this chapter, I will argue that a variety of social forces—especially the scientific recognition of animals as conscious, feeling beings—have created a wide and unstable gulf between humanity's treatment of animals and the belief that animals deserve our protection. Nowhere is this disconnect more severe than our food system. But we will see that the expanding moral circle has eradicated animal cruelty in a wide range of industries and is already banging on the locked doors of factory farms.

The scientific revolution

The scientific revolution kicked off in 1543 with the publication of *On the Revolutions of the Heavenly Spheres*, in which the astronomer Nicolaus Copernicus argued that the sun, not the earth, was the center of celestial orbits—a theory that deeply contradicted the earth-centric religious doctrine of the time.[12]

Copernicus used years of astronomical data to compare the heliocentric and geocentric theories. He determined that the heliocentric model was more consistent with the evidence. If the sun were at the center of the solar system, the appearance of retrograde motion—the apparent backtracking of planets across the night sky some times of the year—could be explained with simple, oval-shaped orbits. If the earth were centered, there would have to be small loops awkwardly inserted into planetary orbits, making it a much less plausible theory. This reasoning is an example of parsimony, a scientific rule-of-thumb that says when we have competing theories that explain our data and evidence equally well, we should select the simplest explanation. With Copernicus and the scientific revolution, parsimony became a hallmark of modern critical thinking.

This heuristic is useful across all areas of decision making, and you've probably used it yourself without realizing it. In medicine, there's an aphorism, "When you hear hoofbeats, think of horses, not zebras." It reminds doctors not to come up with a strange, unusual diagnosis for a patient when a more common one would explain the same symptoms.

Parsimony can just as easily be applied to our understanding of animal minds. As we've accumulated more and more evidence that animals behave like us and even share our neuroanatomy, the simplest explanation is that they are conscious in much the same way we are. If they jump when startled, they probably feel fear. If a puppy rolls over in front of you for a belly scratch, your scratches probably bring her joy.

If it walks like a duck

Thirteen years after the publication of *On the Origin of Species*, the book that popularized the theory of natural selection, Charles Darwin wrote *The Expression of the Emotions in Man and Animals*. He discussed the remarkable similarities among animals, including humans, in their emotional behavior. He writes about the smiling of dogs, the way they show affection like we do, and describes taking his own dog on a walk. He recounts the human-like, fidgety excitement or dejection the dog expressed based on whether Darwin took the right or left fork in a familiar path. To Darwin, this indicated an emotional understanding of the destinations signaled by each direction.[13]

Scientists have even found behavioral evidence of altruism and empathy, two mental capacities long thought to be out of reach for nonhuman animals. For example, elephants will help each other when injured, such as by removing tranquilizer darts and sharp branches. They console nearby anxious elephants by touching them and making soft vocalizations.[14] In laboratory tests, a rat will choose to help a drowning companion over selfishly receiving a tasty piece of chocolate.[15]

Even animals we normally think of as less intelligent have demonstrated remarkable mental capacities. Pigs, for example, have been shown to do much more than snort and roll around in the mud. They form mental categories of different objects in their environment, store long-term memories of their experiences, and even take the perspectives of other pigs.[16] A group of German scientists gave pigs individual names and were able to train them to stand in line and walk up to a feeding station when their names were called.[17]

Let's look closely at the example of grief, a complex emotion many would say only humans can experience. In humans, grief can manifest itself in a number of ways, but it's classically organized into five stages: denial and isolation, anger, bargaining, depression, and acceptance.[18] Individual humans might experience some of these stages, or none. Can animals experience any?

As with most complex animal behaviors, the answer is: yes, in their own way. Scientists have focused on grief behavior in elephants due to their charisma and apparent intelligence. Elephants touch and caress the remains of family members, and place branches and vegetation on the corpses. They treat the site of death differently than other places. When passing through—even years later—they'll sometimes pause in the location for several minutes.[19]

In farmed animals, grief is an inherent component of modern dairy practices. Neuroscientist Oliver Sacks once accompanied the animal behaviorist Temple Grandin to a dairy farm. Sacks recalls:

> When we arrived we could hear a great tumult of bellowing. "They must have separated the calves from the cows this morning," Temple said, and, indeed, this was what happened. We saw one cow outside the stockade, roaming, looking for her calf, and bellowing. "That's not a happy cow," Temple said. "That's one sad, unhappy, upset cow. She wants her baby. Bellowing for it, hunting for it. She'll forget for a while, then start again. It's like grieving, mourning—not much written about it. People don't like to allow them thoughts or feelings."[20]

Without real evidence to the contrary, we shouldn't assume that only mammals or even vertebrates have sophisticated mental capacities like grief. For example, tool use, once thought of as a defining characteristic of humans, or at least primates, has now been recorded in a wide variety of animals. A viral YouTube video even shows an octopus carrying a coconut shell across the ocean floor, then enclosing himself within it, creating a bit of shelter in an otherwise barren environment. Octopuses also stack rocks at the openings of their homes to keep out predators.[21]

Darwin himself went so far as to argue for consciousness in worms, suggesting that they might have at least the most basic level of sentience. "When a worm is suddenly illuminated and dashes like a rabbit into its burrow—to use the expression employed by a friend—we are at first led to look at the action as a reflex one." He goes on to explain that the worm, despite our intuitions, shows behavioral complexity similar to our own, learning and adapting its behavior. He suggests that worms have a spotlight of awareness like we do, shifting their attention depending on the situation. The founder of modern biology concludes that "the comparison here implied between the actions of one of the higher animals and of one so low in the scale as an earth-worm may appear far-fetched; for we thus attribute to the worm attention and some mental power, nevertheless, I can see no reason to doubt the justice of the comparison."[22]

People used to be more skeptical of this sort of evidence, saying that even if animals show complex perception and behavior, we have no way of truly knowing that these behaviors are based in consciousness instead of another mental mechanism. This is known as the "problem of other minds" in philosophy, and it's a tenable philosophical position—but moral action doesn't require certainty. In this case, the simplest explanation is that animals who display behaviors that are driven by consciousness in humans most likely have a similar sort of consciousness. To apply parsimony to this situation, we could use the popular saying: If it walks like a duck, swims like a duck, and quacks like a duck, then it probably is a duck.

Scientific consensus

Over the past few decades, this reasoning has gained major traction in the scientific community, overcoming the Cartesian notion of animals as unfeeling machines. In 1970, scientists began to track killer whales, or orcas, off the coast of British Columbia in Canada and Washington in the US. They discovered that individual whales could be identified with a single photograph of their dorsal fin and upper surface due to variations in tears, skin colorings, and other features. This allowed researchers to track individual whales for the first time, which illuminated the rich social lives of these creatures in matriarchal pods, and—as an indication of how little attention we gave them before—led to the discovery that there are two distinct species of orcas in the Pacific Northwest.[23]

Researchers in this era who drew the obvious conclusion from these results, that orcas were social creatures with complex auditory communication and emotions, were heavily criticized. Raising the profile of whales seemed to threaten the anthropocentric perspective that humans were special in their mental capacities, a viewpoint bolstered by the hunters and businesspeople who exploited these creatures for display and entertainment.

In 1976, animal behaviorist Donald Griffin published a book titled *The Question of Animal Awareness* and received widespread dismissal for even addressing the topic, despite having established himself as a scientific leader with his monumental discovery of echolocation in bats.[24] Today, primatologist Frans de Waal refers to animal cognition as "a term considered an oxymoron until well into the 1980s."[25]

In 1992, biological psychologist Sonja Yoerg still warned young researchers that animal cognition "isn't a project I'd recommend to anyone without tenure."[26] And even in 1995, author Jeffrey Masson wrote in his seminal work *When Elephants Weep*, "There is a tremendous gap between the common sense viewpoint and that of official science on this subject," suggesting that while the layperson agreed with animal consciousness, scientists were still behind the curve.[27]

It wasn't until the twenty-first century that the majority opinion finally shifted. In 2010, psychologist Hal Herzog wrote that 153 of 155 scientists surveyed said that animals could have the conscious experience of pain.[28] In 2012, a number of leading neuroscientists made a definitive statement on the issue with the Cambridge Declaration on Consciousness, noting that "the weight of evidence indicates that humans are not unique in possessing the neurological substrates that generate consciousness."[29]

Today, this new scientific awareness is helping research on the details of animal consciousness flourish. New, fascinating studies are published every year. In 2013, the *New York Times* ran a piece by the psychologist Gregory Berns, "Dogs Are People, Too," describing his research. Berns trained dogs to enter a functional magnetic resonance imaging scanner so he could look inside their brains. Unsurprisingly, he found patterns of activity very similar to those of humans. Although this research is just beginning, Berns called the personhood of dogs his "inescapable conclusion."

This surprisingly recent timeline roughly traces the scientific community's progress away from its dark past, away from the desperate need for human superiority that masqueraded as scientific skepticism. No longer is it standard, or even appropriate, to respond to the evidence of animal consciousness with sweeping dismissals like, "Well, yes, those behaviors seem consistent with consciousness, but we can't really *know* that's what's going on, so it's anthropomorphic and inappropriate to suggest these animals are actually conscious." Members of the scientific community who took this Cartesian view in the past are either changing their minds or moving to the sidelines. (I imagine this is causing them to feel somewhat embarrassed. Then again, even if their behavior seems consistent with feelings of embarrassment, we can't really *know*, can we?)

Popular science

The newfound scientific consensus contributed to the success of *Blackfish*, a 2013 documentary that led to commercial disaster for the animal theme park SeaWorld. The film centered on the story of Tilikum, one of the park's orcas. Tilikum was captured in 1983 near Iceland and transferred to an aquarium in Canada. The young male was confined in a small pool with two older females, leading to significant aggression against Tilikum. Trainers temporarily moved Tilikum into a medical tank on his own, which was even smaller than the already-tiny group enclosure. Orcas are highly social creatures who can roam up to 160 kilometers a day in the wild, so he was eventually returned to his previous enclosure, which didn't bode well for his psychological well-being.[30]

At one point in 1991, when Tilikum was in the pool with the two dominant females, a trainer slipped into the pool and the three whales grabbed and dragged her beneath the surface, forcing her away from a life ring thrown by a bystander and drowning her. Seven years later, after Tilikum had been transferred to SeaWorld in Orlando, Florida, a man who had stayed at the park after closing entered Tilikum's tank for unknown reasons. He was found dead the next morning. Finally, in 2010, Tilikum grabbed another trainer and pulled her under the water, leading her to die from drowning and blunt-force trauma.[31]

Blackfish premiered at the 2013 Sundance Film Festival, and with its testimony from animal behaviorists and SeaWorld employees, made the case that Tilikum's tragic story was only one example of the larger issue of holding any orca in captivity. SeaWorld responded in full force, even creating a new section on its website to address the film, and subsequent backlash fueled further public attention and concern.[32]

Although the impact of *Blackfish* wasn't apparent during its early 2013 premiere, the debate and later airing on CNN prompted a national outcry that continues to this day. SeaWorld's net income dropped 84 percent from the second quarter of 2014 to the second quarter of 2015.[33] Much of the uproar focused on the rich mental lives of the whales, suggesting that the success of *Blackfish* culminated decades of research and advocacy for treating animals as morally important individuals. In January 2017 Tilikum passed away after contracting pneumonia, prompting yet another round of outrage at SeaWorld and marine animal confinement.

The scientific consensus has also contributed to the work of the Nonhuman Rights Project (NhRP), an organization pushing for the recognition of legal personhood for animals, meaning that at least some of them would have the ability to be a plaintiff in court when their rights are violated. Of course, they would have human advocates filing suit on their behalf. NhRP's legal case is "based on the best scientific findings on genetics, intelligence, emotions and social lives of these animals showing they are self-aware, autonomous beings."[34]

One of the NhRP's cases broke new ground on April 20, 2015, when a New York judge issued an order to show cause and writ of habeas corpus on behalf of the chimpanzees Hercules and Leo, who were being held at a medical research facility. This meant that their owner, the State University of New York at Stony Brook, was required to provide legal justification for the decision to keep them captive. The judge eventually dismissed the case, but not because she thought the arguments lacked merit. She felt that her court did not have the authority to make a decision in favor of Hercules and Leo because of the potentially groundbreaking legal ramifications. Despite the dismissal, the university stated that it would no longer conduct scientific studies on Hercules or Leo, and the NhRP entered negotiations to move the chimpanzees to an appropriate sanctuary.[35]

Social forces

Scientific opinion affects, and is affected by, public opinion, but other social factors could help explain humanity's increasing concern for animals. I'll touch on a few of the most interesting ones here because they help us understand how the moral circle will expand in the future, especially in what ways it might reach animals raised for food.

URBANIZATION AND PET OWNERSHIP In the past few centuries, humanity has moved en masse from farms and rural areas to city jobs and urban areas. This migration has been accompanied by a breakdown of traditional social structures. Frequent contact with relatives and tight-knit communities has given way to smaller families and more social isolation. These two factors are correlated with increased pet ownership. As wealth increased and horses made way for cars, city dwellers may have owned pets to show off their wealth or to fill the emotional gaps left by a lack of working animals, relatives, and close friends in the city.

This trend continued well into the twentieth century. From 1967 to 1988, the US dog and cat population as a percentage of the human population rose from 22 percent to almost 50 percent.[36]

Animals make better companions when we see them as individuals instead of just as laborers or useful tools, and this recognition of individuality could be the key factor in the expanding moral circle. Surveys in the past half-century indicate that rural residents, who are more likely to live with animals for their labor instead of companionship, show less affection for animals and less concern about their mistreatment. Rural pet owners instead focus on the animals' "practical and material value."[37] A 2001 study suggested that greater childhood pet attachment was associated with vegetarianism and negative feelings about eating animal products, likely due to increased empathy for animals.[38]

GLOBALIZATION AND FEMINIZATION Concern for animals might also partly result as spillover from our concern for other humans. If we live in a society that rejects human-on-human violence, the similarities to human-on-animal violence might lead to opposition by analogy. Over time, our increasingly globalized society has widened our circle of compassion from family and neighbors, to other villages, then to other cultures and societies.

In *The Better Angels of Our Nature*, psychologist Steven Pinker makes a compelling case that human-on-human violence has overall declined in the course of human history. He discusses several explanations. The majority of them involve mutual benefits and therefore don't seem to apply much to human-on-animal violence, such as the increase in international trade over the centuries that has made war more costly. But two explanations seem quite plausible: cosmopolitanism—also known as globalization, the blending of cultures and perspectives through increasing connections around the world—and feminization, the increasing power of women, who tend to be less violent than men. These seem especially important when we consider psychologically distant animals like those in farms and slaughterhouses. As our society globalizes, we connect with people very different from us through social media, news, traveling, and other mediums. This connection, and the increased compassion as a result of feminization, sets precedent for a general expansion of our moral circle.

As Pinker notes, our increasing concern for animals is a particularly strong reason for optimism that the general trend in violence will continue downwards in the future. Animals have been unable to take political action of their own against their oppressors. Animals are not voiceless, but their cries for help are easier to ignore, and they lack the ability to organize and protest in our political system. This makes them dependent on humanity's ethical choices.

RELIGIOUS TRENDS While some researchers argue that the world is becoming less religious, the main world religions have nonetheless exerted substantial influence over human morality for millennia.[39] In 1888, Mohandas Gandhi moved from his native India to London for law school. While in London, he maintained a vegetarian diet in accordance with his Hindu beliefs and picked up a copy of Henry Salt's book, *A Plea for Vegetarianism*, at a local vegetarian restaurant. Inspired, Gandhi joined the London Vegetarian Society and was elected to its executive committee in 1891. That same year, he graduated, traveling back to India and then on to South Africa. Although no longer in London, Gandhi continued to endorse vegetarianism and strongly oppose animal cruelty. With his rise as a world-famous activist for civil rights and India's independence, vegetarianism in Western society gained social credibility and a strong association with nonviolence and ethical living.

Gandhi is just one manifestation of the religious forces that might have played key roles in the increasing compassion for animals in Western civilization. A religion even more sympathetic to vegetarianism than Hinduism is Jainism, which takes a remarkably broad approach to nonviolence. In addition to not eating animals, its followers avoid plants that grow underground, such as carrots and onions. This is because the harvesting of these plants requires the loss of the plant's life, while harvesting above-ground plants such as corn can allow the plant to continue living.

The first of the Five Precepts of Buddhism asks followers "to undertake the training to avoid taking the life of beings." Taken literally, this might commit a practitioner to vegetarianism because buying and eating animal products requires animals to be killed, although usually the killing is done by someone else. In practice, Buddhist vegetarianism varies by location and denomination, but vegetarianism is generally more prevalent among Buddhists than among nonpractitioners.[40]

As Western societies feel more of a pull from these religions, indicated by the increasing prevalence of practices like yoga and Eastern martial arts, this could bring them closer to adopting ideologies of nonviolence. Even the Beatles were compelled to change, visiting India in 1968 to study meditation under a famous yogi, who served as the band's spiritual advisor.[41] All four musicians went vegetarian at various points in their lives.[42]

Outside of the popularity of specific religions, the movement away from religion, or at least certain forms of it, could be a contributing factor. Religious notions of animals as manifestations of emotions, spirits, and other one-dimensional forces lead us to idolize some species, demonize others, and sometimes do both for the same population. In Native American stories, which were an important part of my own upbringing, the wolf stood as the wise, kind, and honest foil for his trickster coyote brother. The loyalty and camaraderie of wolves is held up as an example for children to emulate, and you never hear of a wolf or coyote who deviated from these predestined roles. In Christian scripture, wolves are sometimes seen as symbols of greed and destruction, challenging noble shepherds and their flocks.[43]

The movement away from these one-dimensional concepts could be enabling us to better appreciate animals' individualities in a way we could not when they were seen as unchanging representations of emotions and other intangible ideas. Arguably, some religions even explicitly diminish the moral value of animals. A famous Bible verse states: "And God said, Let us make man in our image, after our likeness: and let them have dominion over the fish of the sea, and over the fowl of the air, and over the cattle, and over all the earth, and over every creeping thing that creepeth upon the earth."[44]

The word "dominion" is frequently cited as a reason for doing anything we want to animals because they are categorically lower in value than humanity. However, many Christians today see this verse more as a statement of stewardship and responsibility, rather than superiority.[45] In other parts of the Bible, it's suggested that God requires animal sacrifices—though to be fair, it's suggested elsewhere that God dislikes animal sacrifices.[46]

Similar conflicts are seen in other religions. In a perhaps surprising Hindu practice, some followers gather every five years at a Nepalese temple to participate in the brutal sacrifice of hundreds of thousands of animals, hacking away at them with machetes and other cruel implements. Only since 2015 has there been a sign of hope for an end to this outdated violence, as heavy international pressure was put on the devotees to cancel the practice indefinitely. It remains to be seen if animal sacrifice will resume at the temple.[47]

Whether as a result of an alteration or a decrease in religiosity, people are slowly growing to appreciate that animals are complex biological creatures made of the same flesh and blood that we are, not simply manifestations and tools of our own ideologies.

Moving forward

Despite evolving attitudes, we should avoid too much optimism about the expanding moral circle. We are not far removed from previous generations who, could they see where we are now, might resist what we think of as progress, and worry that the moral circle has narrowed in important ways, for example a decreased concern for the preferences of our ancestors, the moral demands of religious deities, or even the natural environment relative to some past cultures. Also, while we usually think of the moral circle as our attitudes, if we instead define it as our behavior, then the advent of factory farming suggests significant *backward* moral progress.

These reasons for pessimism might not be enough to think overall that we have a narrowing moral circle, but they remind us that backsliding is possible, or more likely a stagnation where the circle doesn't fully expand to reach all sentient beings. We need hardworking do-gooders to keep pushing forward. It has become increasingly challenging to maintain the contradiction between our treatment of animals and this moral progress. We subject rats to psychological experimentation to better understand the neurological basis of emotions, trying to convince ourselves that the experiments are ethical because the rats have no emotions. We test painkillers on animals, justifying this by telling ourselves that they don't suffer the way we do. We imagine that dogs lack our own sophisticated awareness of the world, but we still use them for guidance when we lose our own vision.

I will discuss the future of moral circle expansion and associated risks at greater length in the final chapter of this book, but today it remains true that there is great momentum for the expansion of the moral circle to the animal populations, and as I will show in the coming chapters, the moral arc is swiftly closing in on the institution of animal farming.

2. EMPTYING THE CAGES

US FACTORY FARMS—where an estimated 99 percent of farmed animals are kept—are almost as inaccessible to the public as they are inescapable for the animals locked inside.[1] When I started getting more involved in discussions about animal farming, I knew I had to see the inside of one of these facilities for myself. I finally got my opportunity with a rescue team from a California farmed animal sanctuary. The sanctuary was able to convince a handful of farmers to allow them to rescue some of the chickens in their spent flocks. The industry term "spent" is used to describe hens whose reproductive systems are too worn out to be profitable, at which time the animals are killed. These hens were bred to lay eggs, not produce meat, so they can't be sent to the slaughterhouse. Farmers would have to spend money killing the hens and disposing of their bodies; relinquishing them to the rescuers spares the farmers that cost.

The rescue took place at a battery-cage farm in the Central Valley of California in early 2016. Battery cages are thin wire enclosures that are typically so small the hens can't even spread their wings. They're named for the way identical units are stacked end to end, like electrical batteries.[2]

In 2008, California residents passed Proposition 2, a ballot measure in California that added this section to the California Health and Safety Code:

> In addition to other applicable provisions of law, a person shall not tether or confine any covered animal, on a farm, for all or the majority of any day, in a manner that prevents such animal from:

(a) Lying down, standing up, and fully extending his or her limbs;

and

(b) Turning around freely.

Over eight million California residents turned out in favor of Proposition 2, the highest positive turnout for a citizen initiative in the state's history.[3] Many thought this would end the cruel confinement of egg-laying hens in the state. But despite the overwhelming public support for the bill, reform can be slow, and these dreadful cages are still common. In order to comply with the new law, the farm I visited had simply taken out the walls separating the batteries. This means that each hen has more total space in her "cell," but she has more cellmates. Recent investigations have shown that some California farms haven't even made these changes, and there has been only one recorded enforcement of the law in the entire state, which was home to an estimated twelve million egg-laying hens in 2017.[4]

We entered the farm before sunrise so the sleepy birds would be less anxious and easier to handle. We drove through a large metal gate with signs warning KEEP OUT and BIOHAZARD, and passed a dozen sheds with metal roofs and walls of plastic netting. The stench of ammonia was burning our mouths and throats, even outside the barns. We unloaded our transport crates, donned our disposable coveralls, sanitized our boots, and entered the designated shed. From the entrance, we looked down a dozen rows of cages, each approximately 150 feet long. Cages were stacked on each other in two levels, one just above and one just below eye level. On the floor below them was a six-inch-tall, foot-wide mound of feces, which at first looked hard enough to stand on, but the new volunteers got ample warning from veterans that the dry exterior hid a moist, rotting interior. The birds were already agitated from hunger when we came in—the farm didn't waste any money on feeding them before we arrived—but the lights and our presence stressed them out further.

In the absence of rescue, the spent hens face a cruel fate. In normal circumstances, a farmhand pulls a large wheeled box down each row, grabbing the chickens by whatever body part is accessible and ripping them out through the tiny doors of their cages, breaking limbs and tearing skin in the process. The hens are then thrown into the dark, cramped chamber until it's full, and the little air that is left in the box is replaced with carbon dioxide. Injured, confused, and terrified, the chickens die slowly from asphyxiation in a cage of flesh. Hens used for eggs are typically killed at between a year and a half and two years old, a small fraction of the potential lifespan of the chicken's most recent ancestor, red junglefowl. Fortunately, we were there to help—but of the thousands of birds slated to be killed and discarded on this one farm that day, we could save only a few.

We worked in teams to safely transfer the hens from the cages to our transport crates. Some of us reached into the cages to pull out the birds—bruising our arms on the edges of the narrow doors as we carefully pulled out one bird at a time, holding her wings against her body so she didn't flap them and break one on the way out—while other rescuers opened and closed the transport crates. The hens tried desperately to escape our reach, crowding into the corners of the cages and dodging our hands. Some bit us, others cried out in fear when rescuers picked them up, and many were breathing heavily and with open mouths, an indicator of stress. Many birds had wounds from fighting and climbing over each other in the cramped cages, and injuries from incessant contact with the metal cage. Some were missing eyes and had infected cuts from broken wire.

In each cage, you could tell who the dominant birds were and who had been picked on. All the birds were thin, but those who had been bullied were severely emaciated. They were missing many feathers, which other birds had pecked off in agitation. Anxiety had driven some birds to peck off their own feathers. There were many dead birds too, some whose legs and heads hung limp and were beginning to decompose after they had become hopelessly trapped between the wires of their cage walls.

Each rescuer managed the experience differently. Some people's hands shook while they consoled the hens, promising them that they would soon be free of this hellhole. Some quietly let tears roll down their faces while they worked on the task at hand with solemn tenderness. Others were angry and snapped at fellow rescuers for any misstep.

My focus was on diligence and observation. I hoped to absorb as much as I could from the experience, in particular the perspectives of the hens themselves, because I knew my attention to detail could help me better communicate my experience in the future. I needed to remember these stories, not just of the hens who made it out alive to live happily at the sanctuary, but also those who remained on that farm—dead or alive. This harrowing experience helped me put a face, actually more than seven thousand faces in the shed we visited, on the horrors of animal farming.

How did we get here?

The industrial system we now recognize as factory farming emerged in the early twentieth century, when rural farmers struggled to supply enough food for an increasingly urbanized population. In the US, high prices and food shortages led to a nationwide meat boycott in 1910.[5] The start of World War I in 1914 increased the strain on small farms as the US government struggled to feed its troops. Trying to feed a twentieth-century country via a nineteenth-century food system led to protests and riots, such that growing the food supply became a national priority. Farmers and bureaucrats quickly began looking for ways to increase the size and efficiency of the animal agriculture industry.[6]

The first factory farm was probably Mrs. Wilmer Steele's Broiler House in Delaware. Cecile Steele was one of the first farmers to breed chickens specifically for their meat production, starting the genetic division of labor whereby some chickens now lay eggs with rapid frequency while others grow such large breasts they often collapse under their own weight.[7] In 1923, Steele's first flock had five hundred chickens, but due to enormous demand, efficient breeding, and optimized living conditions, she was raising ten thousand per flock by 1926.[8]

Over the next century, to supply the relentless demand for efficiency, large-scale and vertically integrated producers dominated the industry. In 1940, 17 percent of the US population worked in agriculture, compared to 1.5 percent in 2016.[9] Scientific developments like the use of antibiotics in animal feed allowed producers to confine animals in smaller and smaller spaces without a prohibitive increase in disease transmission.

This industrial transformation led to much of the animal suffering we have alluded to so far. Animal farming uses sentient beings as machinery, to produce dairy or eggs, and as raw materials, to produce meat. The industry publication *Hog Farm Management* told its readers in 1973, "Forget the pig is an animal—treat him just like a machine in a factory."[10]

An embedded industry

Farm policy is heavily weighted in favor of animal farming, especially factory farming. As animal advocate Paul Shapiro put it in a *National Review* opinion piece, factory farms "get bailouts when they overproduce, have their most costly business expense (feed) subsidized, get federally supervised dollars to market their products, and even get free research and development that they benefit from but for which they don't pay a cent."[11]

In what *Quartz* called "the US meat industry's wildly successful, 40-year crusade to keep its hold on the American diet," these companies have heavily influenced even the US Dietary Guidelines. In 2015, the scientific advisory committee for the guidelines suggested a shift toward plant-based diets in the interests of public health and sustainability, but the US Department of Health and Human Services and the USDA—which experts agree is heavily influenced by the industry—decided not to use those recommendations and ultimately left out any discussion of reducing meat, dairy, or egg consumption in the published guidelines.

This influence on policy began as soon as US industrial farming emerged. For thirty years after industrial pasteurization equipment became available in 1895, public health advocates pressured local governments to require its use for milk production. Some governments were hesitant to make the changes, but large producers pushed hard for such regulations because smaller milk companies competing with them lacked the funds to quickly purchase and maintain the expensive equipment.[12]

This corporate influence is comparable to that of the gun lobby. While 94 percent of Americans support requiring background checks for all gun buyers and 86 percent support a ban on gun purchases from anyone on the US terrorist watch list, attempts to implement such uncontroversial policies have failed.[13] Experts speculate that the gun industry's power is owed largely to the preemptive measures that were taken to prevent gun regulation in the early 2000s after the industry saw the damage that aggressive regulation had done to the tobacco industry.[14] The difference in outcomes between the movements against the gun and tobacco industries is an important case study for impact-focused activists.

These preemptive measures are similar to antiregulatory laws currently being lobbied for by the animal agriculture industry, deviously called "right to farm" laws. The laws vary in their exact wording, but the recent ones try to curtail future regulation with broad, generic protections. Missouri, for example, actually added one to its constitution in 2014 with a narrow 50.1 percent majority vote.

Critics, including small-scale farmers as well as environmental and animal activists, argue that these sweeping laws could curtail important, beneficial regulations such as preventing factory farms from polluting local groundwater.[15] This coalition is trying to shift the discussion around these laws from the protection of factory farmers implied by the term "right to farm" to the harmful consequences, by calling them "right to harm" laws. At this point, "right to farm" is still more common. Language is a powerful tool, but activists have no monopoly on loading public debate in their favor.

For another example, try filling in the blanks below:

> "Beef: It's ______."
> "The Incredible, Edible ______."
> "Pork: The Other ______."

And, of course,

> "Got ______?"

If you remember those slogans from your childhood and answered "what's for dinner," "Egg," "White Meat," and "Milk," that's because multimillion-dollar government-supervised advertising campaigns worked to drill them into your memory. These campaigns are run by research and promotion organizations funded by companies in their respective industry, such as the American Egg Board (AEB).

The theory behind these "checkoff programs" is that certain

products are largely homogeneous, meaning it's hard to tell one gallon of milk (or cut of beef, case of mushrooms, jar of honey) from another. Since this makes it challenging for individual companies to differentiate and advertise their specific products, the industries partnered with the US government to establish requirements for companies to fund organizations that advertise on behalf of all relevant companies. For example, egg producers with more than seventy-five thousand hens are currently required to "check off" ten cents per case of thirty dozen eggs to the AEB.[16] Government support of these programs—requiring by law that producers participate, even when it's not in their self-interest—gives the industries, most of which are in animal agriculture, significant power in the American marketplace. In the next chapter, I'll discuss how these organizations often overstep their bounds with illegitimate activities, further increasing the power their industries hold over the American dinner plate.

The biggest impact so far

In the fall of 2016, dozens of animal advocates descended on Massachusetts from across the country, joining hundreds of in-state activists, to participate in the "Yes on 3" campaign for the November ballot. I was able to join for the last few days leading up to the vote on the referendum. We went door-to-door discussing the measure, which sought to prohibit the most extreme farmed animal confinement, similar to Prop 2 in California. Most of the conversations were short; the residents just told us they had already voted or were definitely going to vote in favor. In fact, the measure ultimately received an overwhelming 77.7 percent of the vote.[17] Interestingly, almost everyone I spoke to about the initiative brought up how upset and concerned they were by footage they had seen from undercover investigations of animal farms.

This is a common experience for advocates of both plant-based eating and farmed animal welfare, in every situation from handing out provegetarian leaflets to talking with journalists. Almost everyone who cares about the issue does so because of videotaped investigations of animal farms, and the most dedicated, passionate people in the movement are even more likely to be driven by their knowledge of the suffering farmed animals endure—which they learned about through investigations—than other information such as environmental and health concerns. If I had to choose one strategy that has built the most momentum for the movement, it would be the undercover investigations that have exposed and publicized the implications of the modern "machine in a factory" approach to farming animals.

Undercover investigations of animal farms essentially began in the early 1990s, following investigations of animal testing laboratories in the 1980s. Of course, there were earlier exposés, such as Upton Sinclair's work leading to the publication of his novel *The Jungle* in the early 1900s. It seems the first investigation of the modern era was conducted by the pioneering animal rights organization People for the Ethical Treatment of Animals (PETA) in 1983 at a Texas horse exporter, shortly after PETA's famous lab animal investigation in Silver Spring, Maryland, in 1981. More food-industry investigations followed in 1991, when PETA investigated a cow slaughterhouse, a pig slaughterhouse, and a chicken hatchery.[18]

The first modern animal farm investigation, in 1992, exposed the cruelty at Commonwealth Enterprises, a New York farm producing foie gras, the fattened liver of a duck or goose. The major finding was that, contrary to prior claims by the company, the ducks were force-fed. Three times a day, each worker would shove a long, metal pipe down the throats of around five hundred birds. The pipes caused the birds serious pain and resulted in subsequent health issues such as damaged esophagi and pneumonia. This research led to the first US police raid of an animal farm, though reports by activists suggest the foie gras industry successfully pressured the district attorney into dropping the charges. Federal law at that time, and today, excludes farmed animals from standard animal-cruelty protections.[19]

These investigations started receiving major media attention in the late 1990s. A 1998 PETA investigation of a pig-breeding farm led to the first felony indictments ever for cruelty to farmed animals. It "revealed shocking, systematic cruelty from daily beatings of pregnant sows with a wrench and an iron pole to skinning pigs alive and sawing off a conscious animal's legs."[20]

PETA's McCruelty campaign against McDonald's in 1999 and 2000 led to one of the first major corporate commitments to reducing farmed animal suffering, though PETA has since restarted its campaign after McDonald's failed to take continued steps to increase welfare.[21]

While PETA made huge strides for the animal rights movement, impact-focused animal advocates worry that PETA has also done substantial harm to the movement with off-putting campaigns that feature naked women, oversimplify similarities between animal cruelty and historical atrocities such as slavery and the Holocaust, shame overweight Americans, feature goofy animal costumes,[22] and otherwise lead to the public perception that animal protection is a less serious, less important, and anti-intersectional social movement. From PETA's perspective, these gimmicky tactics lead to so much more coverage and exposure for the group and its messages, outweighing the negative reactions. Other advocates disagree. They often struggle to be taken seriously when associations with PETA arise, and they feel the need to bend over backward to convince their audience that they really aren't from *that* animal rights group.[23]

Not only has PETA's hunger for attention ostracized animal protection from other social movements and trivialized animal issues in the eyes of the public and the media, but the organization has attacked the use of research and evidence among animal advocates, presumably because this trend has shifted resources toward other animal nonprofits, especially those working on farmed animal issues.[24] In fact, I would argue that the voracity of PETA and similar nonprofits—their pursuit of attention above all else—has been one of the two biggest mistakes of the farmed animal movement to date. At the end of this chapter, we'll discuss the second issue, which I think is far more ubiquitous and similarly harmful. I do want to be clear that I think these are all only strategic mistakes, and that the vast majority of animal advocates including the leadership of PETA truly have the best interests of the animals in mind.

Another early landmark for the movement, when animal advocates were just beginning to focus on farmed animals, was a 2001 egg farm investigation by Compassion Over Killing (COK). The investigation received national media attention, leading to an interesting quote by the egg producer in the *Washington Post*: "We use normal industry practices. [The investigators'] complaint lies with our industry, not our facility."[25] These days, after over a decade of experience with investigations, industry professionals instead respond to almost every new investigation by condemning the individual facility as a bad apple.

Around this time, a teenager from Ohio founded Mercy For Animals (MFA), which would become quite possibly the most impactful farmed animal organization of the decade. In 1999, when Nathan Runkle was fifteen, a local high school teacher brought a bucket of dead piglets from his farm to class for dissection. One piglet was still alive, so a student who had worked on the teacher's farm implemented the standard practice of "thumping," holding on to the piglet's hind legs and swinging her head into the ground to kill her. The piglet survived that violence despite having a broken skull, and the other students were outraged and attempted to save her. They regrettably failed, but the incident attracted local media attention and inspired Runkle to create MFA later that year.[26]

Since the organization's founding, MFA has relentlessly pursued effectiveness, placing the desire to maximize its impact above all else. MFA's main activity has been conducting and publicizing investigations, such as through provegetarian leafleting and student outreach on college campuses. Its tactical, calculated campaigning has sown a substantial discomfort with factory farming in the US public consciousness. MFA's coverage in almost every leading US media outlet and many leading publications in other countries seems to have steered the global conversation on animal farming.

MFA deliberately limits the number of North American investigations it releases to prevent media oversaturation, exhausting the topic of animal cruelty in the eyes of journalists and readers. Instead of more North American investigations, MFA has prioritized investigations in countries where few or none have been done in the past. The organization is also involved in corporate campaigning for more animal-friendly business practices, which we will come back to later in this chapter.[27]

The other leading organization for farmed animals in the 2000s was the Humane Society of the United States, also due in large part to its commitment to effectiveness. HSUS had for decades been a leader in the protection of pets, wild animals, and other populations, but the organization did little work for farmed animals prior to the early 2000s. Fortunately, the current leadership's focus on impact led them to adjust their strategy to account for the fact that over 99 percent of all domestic animals are those raised for food, and that these animals endure some of the greatest suffering.[28] After their current CEO, Wayne Pacelle, stepped into his role in 2004, HSUS quickly hired Paul Shapiro, the founder of COK, to establish a farm animal protection department.[29]

In 2008, HSUS released a groundbreaking undercover investigation of a California slaughterhouse. It showed, among other cruelties, the use of forklifts to force downer cows (cows so sick or injured they cannot stand or walk) to march to slaughter. Exposure of this practice, which suggested a threat to food safety, especially given the slaughterhouse's involvement with the US school lunch program, led to the largest beef recall in US history.[30] The same year, HSUS led the campaign to pass Prop 2 in California. I should note here that MFA, HSUS, and many of the nonprofits, companies, and individuals we'll discuss in this book are part of effective altruism, the burgeoning social movement and philosophical mind-set on which my research and this book is based.

It's been a delicate struggle for animal protection groups to ensure the media coverage of undercover investigations reflects the important, everyday problems they expose. Journalists and animal agriculture representatives often want to focus on the most explicit, egregious cruelty, like workers punching and kicking animals. This abuse is more headline-grabbing, but is unfortunately easier for the industry to write off as the practices of an errant few. In response, animal protection groups now focus on exposing the rampant cruelty across the industry. MFA says it randomly selects farms to investigate, and other groups have specifically sought out farms with leading humane certifications in order to show that even the animals on those farms—whom many consumers think are happy—still suffer tremendously.[31]

Animal ag's big blunder

Since undercover investigations began, one of the biggest contributors to their success came, inadvertently, from animal agriculture lobbyists. The meat, dairy, and egg industries tried to stop investigations by passing laws in state legislatures to limit the ability to document animal farm operations. The first such law, passed in Iowa in 2011, prohibits undercover audio or visual recording of an animal facility, and possessing or distributing any such record.[32]

In essence, animal ag has sought to punish those who expose abuse, rather than those who commit it.

The public backlash to these "ag-gag" laws, as they are now widely known, began as soon as they were discussed in the mainstream media. Food columnist Mark Bittman popularized the term in a *New York Times* article shortly after the Iowa law was passed. He referenced a Texas investigation from earlier that year which showed calves being hammered to death by workers, animals confined in tiny crates with barely enough room to turn around, pervasive health issues, and the standard industry practice of dehorning calves by burning the budding horns out of their skulls with hot metal cutters.[33]

These laws have been passed in numerous US states, though many have failed, and the Idaho and Utah ag-gag laws were struck down in the courts as unconstitutional.[34] Numerous journalists and public figures have denounced the laws, and that coverage has led many to speak out against eating meat and against the animal agriculture industry as a whole. A psychological experiment even corroborated this effect, suggesting that simply learning about the laws led to decreased trust in farmers and increased support for animal welfare regulation.[35]

One of the leaders in this fight is Will Potter, a thoughtful and soft-spoken investigative journalist and author of *Green Is the New Red*, a book about the government and industry repression of environmental and animal rights activists, including the use of the provocative label "eco-terrorist" to disparage peaceful activists.

In April 2013, Potter debated Emily Meredith, director of communications for the Animal Agriculture Alliance, on *Democracy Now!*, an independent progressive news program. Potter pointed out that the ag-gag laws are opposed by many groups outside of animal protection, such as the American Civil Liberties Union and National Press Photographers Association. The reasoning of these groups is straightforward: we wouldn't accept laws silencing whistleblowers in any other commercial workplace—senior living facilities, day-care centers, textile factories—so why give animal farms and slaughterhouses special treatment?

The best argument Meredith could muster was that the videos caused damage to farmers and their reputations because they were "harsh criticism." She felt the investigators were incriminating farmers by releasing the videos directly to the media without giving them an opportunity to defend themselves or take corrective action, to which Potter responded by noting that advocates consistently reported the findings of their investigations to authorities, resulting in numerous felony charges.

A frequent rebuttal from Meredith and other industry representatives is that animals would be better served if witnesses reported incidents of cruelty immediately, so the animals could be helped as soon as possible. While this might sound like a tenable position, animal-cruelty allegations simply aren't taken seriously unless there is a lengthy record of abuse, and given the lack of any meaningful welfare laws for animals on farms, documenting abuse is necessary to expose conditions on animal farms and generate public pressure to see justice done for the animals.

In June 2013, after this debate and another on CNN where even the moderator appeared to take the side of animal advocates, Meredith appeared on *The Daily Show*. Among other provocative, humorous questions, she was asked, "So let me get this straight, you're protecting the animals from the people that are trying to protect the animals?" She answered with a clear yes, perhaps not fully understanding the loaded question. Later in the interview, Meredith raised a concern about the activists using their videos for fundraising, and the interviewer asks, "So it's the activists that are connected to the bottom line, not the farmers?" Meredith responds, "Yes, the animal activist industry is a huge business." Following this statement a graph showing food-company revenue dwarfing that of animal-protection organizations appeared on-screen. I'm not sure if Meredith got the joke.

Since 2013, the industry quietly continued to push ag-gag laws through state legislatures, despite public opinion. In August 2013, a policy representative from the National Pork Producers Council said, "We did a study of coverage of 'ag-gag' laws that found that 99 percent of the stories about it were negative."[36]

Stepping outside the echo chamber

Animal activism is no longer on the fringes. Farmed animal activists at animal conferences often begin slideshow presentations with pictures of younger versions of themselves. True to stereotypes, many movement leaders got their start in the punk scene, with dreadlocks, pierced noses, or my favorite, the ten-inch yellow mohawk of David Coman-Hidy, who is now executive director of the Humane League, a leading farmed animal protection nonprofit.[37]

While there's nothing wrong with that style, many of those activists have discovered that business casual allows them to reach new audiences—they can even enter the corporate boardroom, where so many important decisions for animals are made. These activists have stepped off the streets, tucked the public outrage sparked by undercover investigations into manila folders, and slammed them onto the desks of CEOs with demands for concrete policy change.

The victories they've achieved help animals in truly staggering numbers. For example, in 2015, the year these reforms started picking up steam, Mercy For Animals ran successful campaigns for higher welfare policies at top food companies that impact an estimated seven million farmed animals annually. Most of the animals affected are egg-laying hens, who will now live in cage-free sheds instead of constricting battery cages. Other policies banned cruelties like tail docking, the standard industry practice of cutting off dairy calves' tails with no anesthetic.

That seven million figure comes from calculations by Animal Charity Evaluators (ACE), the research organization where I previously served as chair of the board of directors and then as a full-time researcher. We took the number of affected animals used by these companies, such as dairy cows if it was a tail-docking policy; accounted for our best estimate of the proportion of impact MFA was responsible for, since other organizations and pressures also contribute to the success of these campaigns; and reduced the estimated impact for those corporations publicizing welfare aspirations rather than actual policy.

Using a similar methodology in 2016, ACE's estimate of MFA's corporate impact skyrocketed to 150 million animals, due to both rapid progress in cage-free egg policies and the introduction of reforms for chickens raised for meat. The latter reforms included less cruel slaughter methods, the addition of windows and perches to the closed sheds the birds spend most of their lives in, and the use of higher-welfare chicken breeds that have fewer health issues. MFA's 2016 corporate campaigns budget, including all overhead costs, was just over $1.1 million, so these campaigns appear tremendously cost-effective.[38]

Leah Garces is the executive director of the US branch of Compassion in World Farming (CIWF), a UK-based farmed animal protection organization leading many of these corporate campaigns. Garces told me she got her start helping animals as a teenager in the 1990s, going vegetarian because some of her friends were, including her boyfriend, who is now her husband. In college, as her concern for animals grew, she defaulted to the veterinary medicine career path, but a thoughtful professor at the University of Florida suggested to her that veterinarians were like plumbers—fixing individual issues as they come up, but not improving the system itself.

Garces has been able to build invaluable coalitions to help farmed animals at the national and global scale. The movement stretches across a range of ideologies and intensities. Many advocates who strongly oppose animal farming are more comfortable being as far away as possible from company executives and factory farmers—the people they see as causing intense suffering to billions of animals. But because corporations have so much control over the lives of these animals, advocates who successfully negotiate with them directly can effect big changes.

Garces, a unifier nonpareil, told me about her visit with Craig Watts, a fed-up factory poultry farmer working with Perdue, one of the nation's largest chicken companies. She hoped Watts would join CIWF's campaign against factory farming, but as she was driving out to his farm with a film crew, Garces worried about the possibility of an ambush by farmers because Watts sent her some joking texts about his industry friends. She decided to persist, however, and her courage paid off: the factory farmer's denouncement of Perdue's practices achieved major news coverage.

Now, there has been some debate about how much welfare reforms, particularly cage-free egg policies, actually benefit the animals. Unfortunately, in some ways cage-free sheds are even worse than battery cages. For example, the ground of cage-free sheds is typically covered in feces, which the birds root through and stir up as they walk around, resulting in poor air quality and respiratory illnesses. And the large number of hens—typically thousands, sometimes over ten thousand—crowded together in one shed makes it difficult for them to establish a pecking order. Because hens can't sort out their place in the social hierarchy, they end up constantly fighting. Hens also injure each other with forceful pecking, and some die from what is effectively cannibalism. This harmful behavior is thought to be driven by the lack of appropriate pecking substrates like litter or dirt for young chickens in a barren environment, meaning the birds direct their natural pecking drive at the only available option, other chickens. It spreads more easily in a cage-free environment due to the large number of individuals interacting with and learning from each other. This abnormal behavior does not develop in small flocks of free-range or backyard birds with plenty of space and an interesting, complex outdoor environment.[39]

Ultimately, researchers mostly agree that the direct benefits of cage-free reforms—mainly the ability of hens to express some semblance of natural behaviors like perching, dustbathing, and just walking around—outweigh the downsides. But the downsides remind us not to set our sights too narrowly on short-term reduction of suffering.

Another skeptical question we can ask about the impact of these reforms: Are activists really the ones causing these policy changes, or are they just taking responsibility for actions the food companies would take anyway? Well, you don't have to take the word of activists—take it from Chad Gregory, president of the United Egg

Producers, an industry cooperative representing around 95 percent of US egg production.[40] Here's a section of a March 2016 article in *Feedstuffs*, the largest US animal agriculture news publication:

> As for what's causing the rapid rate of announcements, [Gregory] said HSUS and The Humane League have been "harassing and bombarding" every retailer, every food manufacturer, every food company there is with a "very aggressive" cage-free campaign.
>
> "They're the ones that have been driving this; there is no question about it," he added.[41]

Further evidence of advocate responsibility comes from the timing of announcements: companies rarely make policy changes except shortly after a round of advocate pressure. In private conversations with researchers, industry executives have confirmed the impact of advocate pressure.[42]

The skeptical reader might still worry that welfare reforms don't reduce the size of the animal agriculture industry. They might even increase the industry's long-term success by alleviating the concerns of consumers who are complacent in the belief that they can buy the products derived from those animals guilt-free.

I was worried about such complacency in the early days of my research, and I still actively look for ways to prevent it. However, animal advocates have debated the longer-term effects of welfare reforms for years, and I've come out convinced that the positive indirect effects outweigh the negative indirect effects. In particular, I think pushing for these policies generates more momentum against animal farming than complacency with the current system.[43]

Let's dive into this a little more and outline the three main reasons to expect momentum over complacency. First, current evidence suggests that most successful social movements have also taken this sort of incremental approach, even with their bigger goals in mind. British advocates fighting slavery, for instance, initially focused on abolishing not the whole establishment of slavery, but the trade that supplied it, with the intention of starving the broader institution to ultimately take it down. They also worked to ameliorate the condition of existing slaves. In fact, the slave industry's refusal to make more than very minor concessions—and the difficulty of even making and enforcing changes as paltry as requirements for subsistence standards for food, clothing, and shelter—seems to have spurred the public outrage that fueled the final push for abolition. We could see a similar process of attempted reforms, failure to reform, then a push for abolition with modern industries like animal farming.[44]

Second, there is some evidence, both theoretical and empirical, that welfare reforms might reduce total production through their effect on price. Namely, it tends to be more expensive to treat animals better; if higher welfare practices were profitable, the industry would likely already be implementing them. Empirical data points supporting this theory include the rises in egg and bacon prices following the European Union's passage of anticonfinement legislation, which banned battery cages and most uses of gestation crates—metal enclosures for pregnant pigs so small the animals can't turn around—and a similar rise in egg prices and subsequent drop in production by 35 percent due to the passage of Proposition 2 in California.[45]

There's an interesting tension here between making these reforms acceptable to the mainstream and having the maximum ethical benefits. Many advocates try to downplay these possible price increases to make the reforms seem more palatable to meat-eating consumers, but I'm happy to endorse the increases. We pay more for more ethical standards in other industries, and animal farming should be no exception. Price increases will also, all else being equal, reduce overall consumption of animal-based foods and increase consumption of healthy, ethical, affordable plant-based foods. I don't hesitate to say that this is a positive side effect.

Finally, consider how these reforms inspire the activists and donors who see real, significant changes happening because of their contributions. These incremental incentives are easy to forget about, but if farmed animal advocates only fought for major policy change to decrease or eliminate the production of animal products, the lack of tangible successes in the meantime would likely discourage many of them.

Welfare victories establish precedent for more than just the movement and industry. They signal to media, intellectuals, and the public that farmed animals matter and that farmed animal advocates are a powerful force, worthy of press coverage and serious consideration. Just as individuals work to be consistent in their behavior, so does society: taking one action to help farmed animals strengthens our identity as a society that cares for farmed animals, increasing the likelihood of further welfare reforms and the adoption of animal-free foods. This is backed up by some limited experimental evidence from three small-scale experiments in which participants who were given information about welfare reforms took greater interest in reducing their animal-product consumption than participants who were given unrelated information.[46]

Victories also build relationships. A farmed animal advocate who pens an op-ed on the benefits of cage-free housing can later pitch an article on animal-free food to the same editor using that established trust from having worked together in the past. A nonprofit that works with a food company on welfare reforms can later work with them on animal-free food adoption.

There is also evidence of a correlation between countries with higher welfare standards and higher rates of vegetarianism. This evidence should be qualified by the likely influence of outside variables, such as a more embedded general concern for animals, which probably leads to both higher welfare standards and higher vegetarianism rates. Besides cross-country data, similar correlations have been found within the US and the Netherlands, where people who favor "higher welfare" animal products are more inclined toward vegetarianism.[47]

Vegans, vegetarians, and reducetarians

When investigators and other advocates share videos and otherwise engage with the public, they need to include an "ask," or request—some action an inspired viewer can take if they want to help end the animal cruelty, environmental damage, or other harm they've seen. This activism can be very compelling. A recent advance has been the use of virtual reality headsets to immerse the audience in the life of a farmed animal. This tactic has been pioneered by Animal Equality; one video takes the audience through a chicken slaughterhouse via a camera hanging between two chickens, as a worker sharpens his knives in front of them, ready to slice the throat of each passer-by.[48] For the past three decades, the movement's ask for viewers eager to take action has almost exclusively been for individual diet change, either for the viewer to reduce their meat consumption or leave animals off their plate entirely.

How successful has this been? The data here is limited. The few surveys that have been done over the years, mostly in the US, have varied in their methodology. A 1978 poll found that 1.2 percent of those surveyed self-identified as vegetarian, and a 1992 poll found 7 percent self-identification.[49] After 1992, the best polls we have are those conducted by the Vegetarian Resource Group, which asked respondents whether they "never eat" certain products and counted as vegetarian those who checked off each category of animal flesh: red meat, poultry, and seafood. The results were 0.7 percent in 1994, 1 percent in 1997, 2.5 percent in 2000, 2.8 percent in 2003, 2.3 percent in 2006, 3.4 percent in 2009, 5 percent in 2011, 4 percent in 2012, and 3.4 percent in 2015.[50] Around one-third to one-half of self-reported vegetarians in each poll indicated they consumed no animal products.

We also have the Gallup poll conducted in 1999, 2001, and 2012 that asked whether people self-identified as vegetarian. It found that 6 percent, 6 percent, and 5 percent did, respectively.[51] This poll data is only weak evidence because of the statistical margin of error in these polls, but overall, this suggests that there was a significant increase in the number of self-reported vegetarians sometime in the late 1990s or 2000s. Anecdotally, veteran activists basically agree with this trend, and it syncs up with the release of investigations and new plant-based products over the past few decades.

We have little data for other countries, but polls mostly suggest figures between 2 percent and 12 percent, with the clear outlier of India at 29 percent, largely due to religious vegetarianism.[52]

Survey results indicate around 54 percent of Americans are self-identified reducetarians, meaning they are "currently trying to consume fewer animal-based foods (meat, dairy, and/or eggs) and more plant-based foods (fruits, grains, beans, and/or vegetables)."[53] We also have some data for the total number of animals slaughtered in the US, which suggests an increase until 2008 and a stabilization thereafter, which is promising given the continued growth of the human population.[54]

The base of this pyramid of concern for farmed animals is even wider. A 2014 US survey found that 93 percent of respondents felt it was "very important" to buy their food from humane sources.[55] Eighty-seven percent believe "farmed animals have roughly the same ability to feel pain and discomfort as humans." And an astounding 47 percent of US adults say in a survey that they support the seemingly radical policy change of "a ban on slaughterhouses."[56]

As we conclude our discussion of the current state of the movement and move on to future directions, I want to note my concern with this heavy focus of the movement on individual diet change, which has made outrage at animal farming virtually synonymous with vegetarianism and veganism. In my view, this has alienated many potential supporters and overall has been one of the two biggest mistakes of the movement, along with the gimmicks and attention hunger of PETA and similar groups discussed earlier. Fortunately, we will see throughout this book that the movement is shifting toward a focus on institutional change, targeting companies, social groups, and society at large with impressive results.

3. THE RISE OF VEGAN TECH

JOSH BALK AND JOSH TETRICK—the Joshes—defy vegan stereotypes. In college, Balk played baseball and Tetrick played football. They're imposing, with big smiles and charming wits. When you ask them how their work is going, they compliment everyone but themselves: their team, their customers, and especially you.

In other words, you'll want to buy anything they try to sell you.

"Handsome Creek"

The Joshes founded Hampton Creek (now known as JUST) in 2011. I first visited their San Francisco office in 2015, where my colleagues and I were welcomed by a tongue-lolling golden retriever named Jake who might have been officially employed as the office greeter. We sat down on the couch near the front of their office as we waited for our tour. It was the prototypical Bay Area tech startup workplace, with Apple computers, an open floor plan, and seemingly not one person over the age of forty. Tetrick was sitting on a bar stool chatting with some well-dressed investors, whom I later learned represented Li Ka-shing, a Hampton Creek investor and one of the richest people in Asia.

Because we were nonprofit workers, our tour was guided by Balk. Balk has actually never been employed by Hampton Creek, instead opting for nonprofit work, primarily with HSUS, where he's currently vice president of farm animal protection, having replaced Paul Shapiro.

Before working with HSUS, Balk worked as an undercover investigator. He told us about his investigation of a chicken slaughterhouse around 2004, where he took a job in order to secretly videotape conditions at the facility. On the slaughter line, workers grabbed chickens by their legs to pull them out of transport crates and held them at eye level to shackle them upside down from a metal cable that pulled the birds through the slaughter process. First it dragged them through an electrically charged water bath to immobilize them, then it slid them across a spinning metal blade to slit their throats while they were still conscious, and finally it dropped them into a pool of scalding hot water for defeathering. At the last stage, many birds were still alive because the first two steps did not reliably kill them. Balk's smile slipped away when he told us this.

The repetitive labor of working the slaughter line is often physically and mentally stressful and degrading for slaughterhouse workers. Balk had these problems and more as someone who appreciated the moral personhood of each individual animal he was sending to the 150°F water. Balk stayed motivated—and sane—by knowing he was capturing footage of this cruelty that would potentially help millions of animals avoid this fate once it was released to the public.

In those early days of the farmed animal protection movement, the investigations were nowhere near as sophisticated as they are today. Balk wore a large, cumbersome recording device under his shirt. On the slaughter line, workers would tap each other on the abdomen to indicate it was time to switch line positions, but since investigations weren't as well-known back then, Balk managed to evade suspicion despite the strange object—perhaps his coworkers thought it was a medical device.

On his last day at the facility, he became bolder. He openly displayed a camera to get some higher-quality photos of the birds and the facility. The head of security caught Balk taking pictures, but let him go after Balk showed his worker ID, warning him that he didn't want to see the photos on *60 Minutes*. Balk even talked the security guard into taking a photo of him next to the birds. As we will see, this audacity has also driven bold—and very successful—business strategy at Hampton Creek.

When he told us about his transition from investigator to policy at HSUS, his smile emerged again. Now he puts on a business suit and meets with animal agriculture executives to persuade them to adopt higher welfare policies. Balk is nothing if not a dedicated pragmatist. In his fifteen years of advocacy work, he has never taken a sick day and has only missed work for one vacation, a few funerals, and to volunteer on election day for various candidates—including his dad—five times. Despite this relentless work ethic, or perhaps because of it, Balk's smile rarely fades.

One of Balk's key realizations at HSUS was that the executives at big companies, despite their reputations as selfish scrooges unconcerned with ethics, are typically just as uncomfortable with animal cruelty as the rest of us. According to Balk, they see current business practices as a necessary evil, but an evil nonetheless. Of course, these savvy business leaders were surely to some extent just trying to placate HSUS, but the evidence they showed of genuine concern nagged at Balk constantly in the days before Hampton Creek.

He told me that he remembered one meeting in particular with General Mills where the executives clearly wanted their company to adopt a cage-free policy, but they felt hopelessly constrained by the high financial cost of transitioning. After wrapping up his visit, he found himself on a grounded plane at the Minneapolis–Saint Paul International Airport, delayed by snow. He felt frustrated at the inability to change corporate policies, thinking: "General Mills and most any other company would be happy to use non-animal-based products and ingredients as long as the taste was the same or better, and the cost was more affordable. Well, [a company doing that] doesn't exist yet so why [don't] I just start one?"

When he got home, he called Tetrick—they had been friends since they were fifteen years old—and pitched the idea.

After graduating from college in 2004, Tetrick worked as a Fulbright Scholar teaching schoolchildren in low-income countries. Through this lens, he developed an interest in both the power of business and food to help others. He went to law school and worked on climate change business law until 2009, when he wrote a column titled "You Can Save the Planet" for the *Richmond Times-Dispatch*, in which he included one line on animal agriculture: "Seventy billion animals—about the number of humans who've lived in all of history—suffer from cruel and inhumane treatment inside factory farm walls." For this single factual statement, Tetrick lost his job because the largest pig company in the world, Smithfield, was one of his firm's clients.

Hampton Creek has never tried to develop a product targeted at vegans or at ethically minded consumers who can afford niche products. The Joshes simply wanted to take advantage of the inefficiency of feeding plants to animals then feeding those animals to humans, and instead make plant products directly with lower costs. They wouldn't have to persuade any corporate executives to choose ethics over profits; they would just count on those executives' desires to pay the lowest price possible. Eggs seemed like a great place to start given their ubiquity in food products that are relatively easy to replicate, like mayonnaise, and the particularly devastating animal-welfare harm of egg production.

The first step in this potentially revolutionary endeavor was to ensure that their plant products could satisfy all the culinary needs of consumers and food professionals. Could their plant-based eggs make a soufflé rise? Add moisture and binding to a cake? Emulsify mayonnaise? Scramble in a pan for breakfast? To solve this, Tetrick recruited Johan Boot, former R&D director at Unilever;[1] Joshua Klein, a protein expert who previously worked on a gene therapy for HIV in a multimillion-dollar research project supported by the Bill and Melinda Gates Foundation;[2] and Chris Jones, former *Top Chef* contestant and former executive chef at Moto, a world-famous molecular gastronomy restaurant. Jones actually brought an entire crew with him from Moto to support Hampton Creek's culinary research. The San Francisco startup scene is known for attracting big talent, but this was exceptional, and all for a company focusing on egg-free mayonnaise.

The key to Hampton Creek's research has been the trial-and-error testing of a huge database of plants, and the success of that testing has come down to individual species with unique properties, like mung bean protein, which was found to coagulate with heat, making it a prime candidate for plant-based scrambled eggs.

Hampton Creek's approach reminds many advocates of the technological revolutions that disrupted other animal-based industries like whale oil, replaced by kerosene and vegetable oil, and horse carriages, replaced by trains and automobiles. In 1898, the first international urban-planning conference convened in New York City, not to discuss crime or housing, but horse shit, of which there was more than four million pounds deposited in the city per day. City officials could find no solution to the expense, disease, and traffic gridlock resulting from not just manure, but horse carcasses, which were allowed to decompose in the streets so they could be more easily sawed apart and moved. The urban planners gave up three days into their planned ten-day conference, but by 1912, internal combustion technology led to more cars than horses in New York City for the first time, sparing human and animal residents from misery.[3]

Controversy

As Hampton Creek perfected its technology and secured significant investments from Silicon Valley titans, the American egg industry took notice. First, in 2015 the food giant Unilever sued Hampton Creek on the grounds that Just Mayo was an inappropriate name for an eggless product. Unilever expected the new company to back down like most small companies do when threatened, but the Joshes were determined and Hampton Creek stood firm. Balk's experience campaigning against big food companies had taught him what a steadfast underdog can do in this industry.

This ignited a public relations fiasco for Unilever as the media cast it as a faceless Goliath attacking a small, ethically minded David, with a friendly, down-to-earth CEO. The coverage in the first week alone gave Hampton Creek $21 million worth of free, overwhelmingly positive media coverage. Unilever eventually dropped the lawsuit.

In August 2015, the FDA got involved. The regulatory agency sent Hampton Creek a warning letter based on mayonnaise's "standards of identity," intended to keep food companies from using deceptive labels that mislead consumers.[4] Tetrick fought back again, earning more consumer recognition. Hampton Creek eventually had to make minor changes to the label: it shrank the size of the logo—an egg silhouette with a plant growing in the bottom, increased the size of descriptive terms like "egg-free," and added the phrase "spread and dressing."[5]

Yet the mayo war continued. Just a month later, and a month before I visited Hampton Creek's office, the *Guardian* announced findings that the American Egg Board (AEB) had plotted against the startup. The AEB's marketing tactics aren't supposed to include attacks on other industries, but the AEB was nonetheless caught scheming to keep Just Mayo out of the retailer Whole Foods. Emails publicized by the *Guardian* even included jokes of violence against Tetrick, one asking, "Can we pool our money and put a hit on him?" One official offered "to contact some of [his] old buddies in Brooklyn to pay Mr. Tetrick a visit."[6]

Besides sparking another media frenzy—again highly positive toward Hampton Creek—this scandal gave serious credibility to the startup, as it was clearly seen as a threat by the formidable egg industry. This was an encouraging sign to everyone working to defeat animal agriculture. The day I visited the office, news broke that the AEB CEO had resigned after the scandal, leading to yet another round of free, positive press coverage, and showing the power of activists and mission-driven companies to make a significant impact.

Unfortunately, in August 2016, Hampton Creek experienced a round of negative media attention. *Bloomberg* criticized a 2014 buyback program where contractors were paid by Hampton Creek to purchase Just Mayo from stores. According to Tetrick, this was a quality control program—meaning contractors bought the product from a variety of locations to ensure it had the right color, taste, undamaged packaging, and other features—with a budget of around $77,000, a figure Hampton Creek investors say was verified in an audit by professional services firm PricewaterhouseCoopers, lending a little more credibility to their side of the story.

Bloomberg alleged that the real figure for just one month of this program, July 2014, was $510,000, a significant figure compared to $472,000 in company sales that month.[7] The media outlet cited statements and documents from former Hampton Creek employees, but I haven't personally been able to find any third-party verification. The US Securities and Exchange Commission took the report seriously, opening a preliminary inquiry into the buyback program, though this inquiry has since been closed with no finding of wrongdoing.[8]

In a CNBC interview, Tetrick called for viewers to think about the broader context of the allegations. CNBC then displayed a graph showing the total sales of Just Mayo—a sharply increasing line from

December 2013 to June 2016—towering over a tiny horizontal line that included a brief vibration with the sarcastic label "Great Mayo Buyout of 2014."[9]

Still, a buyback program with the sole focus of boosting sales isn't unusual for an ambitious startup. In their book *The Method Method*, the founders of the eco-friendly cleaning products company Method—which includes Adam Lowry, cofounder of the plant-based dairy company Ripple, discussed in the next chapter—wrote about buying their own products at Target and passing them out for free in parking lots. Walt Disney had his friends call theaters to ask about showings of Mickey Mouse cartoons, and if they weren't being shown, to ask the theater why not.[10] Sara Blakely, founder of the underwear company Spanx and the youngest self-made female billionaire in American history, paid people to buy her products and even went into department stores to move her product closer to the cash registers.[11] All of these examples were seen as funny, intimate stories of ambitious people trying to launch small businesses, suggesting that the negative media coverage of Hampton Creek's program was driven more by the attention a story about an ethics-driven company with shady business practices would generate.

Toward the end of his interview, Tetrick acknowledged that getting stores excited about Hampton Creek's product was a "secondary purpose" of the buyback program. In another interview, Tetrick further admitted that there were some purchases outside the company's budgeted $77,000 to check on other store features like pricing and shelf positioning.[12]

Late in August 2016, Tetrick told the *New York Times* that Hampton Creek was still operating at a loss, five years after the company's founding.[13] But this is still not that unusual for a tech startup. In fact, Hampton Creek's approach of keeping prices low while it grows to reach economies of scale seems like the right strategy, especially because it is selling to food service companies and other institutions and has a good understanding of how to produce more cheaply at a larger scale. This is common practice in today's tech industry with companies like Uber.[14]

Overall, I strongly agree with Tetrick that we need to focus on the broader context. The egg industry is surely guilty of far greater ethical transgressions than Hampton Creek, even if we take all of *Bloomberg*'s allegations as true. Journalists pounce on any controversy surrounding well-known companies like Hampton Creek, especially if those companies pride themselves on ethics. We should constantly remind ourselves that the alternative to these new companies is business as usual, a reality virtually nobody is happy with.

That being said, if we read between the lines of Tetrick's statements, I do think Hampton Creek's decision to have employees—not just excited customers—call retailers to express interest in its product was misleading. This might be standard practice for startups, but companies innovating for a more ethical food system should hold themselves to a higher standard. Beyond the direct impact of shady practices, there's a risk of harm to the reputation of the animal-free food industry. Even if a company like Hampton Creek is overall a force for good, one mistake can taint public views of other companies with similar goals, and reduce the chance of all of us succeeding in building a better food system. Despite its mistakes, Hampton Creek's transparency is laudable. The company owned up to a lot, such as Tetrick dating an employee early in the company's history and a failure to correctly label an ingredient "lemon juice concentrate" instead of "lemon juice"—which it promptly corrected.[15] As consumers and observers of an emerging ethical industry, we should be careful not to give up on imperfect companies that combat the systemic evils of the food industry.

This controversy also suggests what a big influence journalists have over new technologies, able to spark public outcry or fanfare on a whim. We'll see these effects again when it comes to how journalists label animal-free food, as their incentive to drive clicks might lead to an emphasis on misleading, negative terms like "lab-grown meat."

From tofu to vegan turkey

The first vegetarian "meats" were peashooters compared to the B83 1.2 megaton nuclear bomb of the modern plant-based meats. According to *History of Meat Alternatives (965 CE to 2014)*, a book that introduces itself as "the single most current and useful source of information on this subject," the first reference to plant-based food that mimicked animal flesh was a description of the canonical vegetarian protein tofu in AD 965. It's said that the magistrate of the Chinese city Qing Yang encouraged tofu consumption as a more frugal alternative to animal flesh, referring to it as "mock lamb chops" and "the vice mayor's mutton."

Nobody's sure how tofu was first made. It might have resulted serendipitously from an accidental seasoning of soybean soup with unrefined sea salt containing magnesium chloride, a coagulant still used in modern tofu production.[16] This would have led solid, white curds to form, i.e., tofu. This is still by far the most popular vegetarian meat, remaining a staple even for nonvegetarians in China and Japan.[17] In the next nine centuries, we find historical references to seitan—made from wheat gluten with a chewy, often sinewy texture that makes it more meaty than tofu; yuba—also known as tofu skin or bean curd skin, which is the solid sheets that form on top of boiled soymilk; and tempeh—a fermented soybean product originating in Indonesia with a rich, nutty taste.[18]

In 1852 we see the first reference to vegetarian meat in Western civilization. This food was less creative, just a sausage-like mixture made by squeezing together finely chopped turnips and beets. More sophisticated foods developed by the 1890s as cooks combined nuts with wheat gluten to make seitan. The vegetarian community at this time, many of them Seventh-Day Adventists, continued developing nut-wheat products over the next few decades. The first recorded veggie burger appeared in 1939, containing soybeans, wheat gluten, peanuts, tomato paste, and seasoning.[19]

The VegeBurger, arguably the progenitor of all modern plant-based burgers, was developed in the late 1960s, where it was made at Seed, a London macrobiotic restaurant serving the likes of John Lennon and Yoko Ono. It was made from a blend of sesame, soy, oats, and wheat gluten. It was sold in grocery stores as a dry mix that consumers could rehydrate, form into patties, and fry or grill.[20] We lack reliable market data for this time period, but it seems vegetarian food rose quickly in popularity astride the counterculture movement of the 1960s.

The most famous US vegetarian meat brand is probably Tofurky, made by the company Turtle Island Foods.[21] In true hippie fashion, the company named itself Turtle Island as a reference to Native American folklore in which the world was flooded and a giant turtle carried the land animals on his back, becoming the North American continent.[22] I remember this story from my own childhood. Like the hippies of the 1960s, it helped me connect with and feel more compassion for the natural world and its inhabitants.

Turtle Island released its famous Thanksgiving roast (popularly known by the generic term "tofurkey") product in 1995, complete with stuffing in the center. With its first five hundred tofurkeys sent out by mail-order, Tofurky solicited feedback with prepaid postcards. The product was a hit. One customer wrote, "I have been waiting 20 years for this product. Finally I am not a second class citizen at Thanksgiving anymore!" Thanks to media coverage on NPR and the *Today* show that year, Tofurky helped catapult vegetarianism into mainstream discourse.[23]

Plant-based meat

When I met Oliver Zahn in 2015, he was director of the Center for Cosmological Physics at the University of California, Berkeley. Zahn was a fellow member of the local effective altruism community, a social movement and philosophy based on trying to maximize one's positive impact on the world. In July 2016 I helped the German-born scientist and his family pack up some of their possessions as they prepared to move out of their California home. By this time, Zahn had transitioned to apply his expertise to a mission-driven startup, working as chief data scientist at Impossible Foods, one of the most famous animal-free food companies today. Instead of mapping out the stars to trace the origin of the universe, he was now mapping out plant ingredients to build an animal-free food system.

As the team helping with the move grew hungry, Zahn pan-fried burgers for us with some frozen plant-based beef that he had brought home from work. This was my first encounter with an Impossible Burger, what was already being referred to in media outlets as the food of the future. The pink, raw patties were visually unmistakable from animal-based ground beef. The first thing I noticed in the cooking process was the distinctive color change to the grayish-brown color of a typical beef burger. The Impossible Burger's outside char was a little crispier than that of beef, and overall the patty looked a little dry, but I worried that I was just imagining differences because I expected the product to be imperfect in some way.

When it was my turn to try one, I opened my mouth wide and posed for a picture, then took my first bite. To be honest, I actually couldn't distinguish what the patty tasted like apart from the bread, lettuce, and condiments packed together, so I took out the patty and bit into it alone. I was blown away by the complex, rich flavors and truly meaty texture. I've enjoyed plenty of ersatz burgers, because I lost my taste for meat after being vegetarian for so long, but this is the first one I tried that captured the unique culinary experience of animal flesh. The patty was a bit thin and dry, but overall, I couldn't complain, and I knew that those issues could be fixed in future iterations. As a meat eater who was enjoying the Impossible Burger along with me said, "I wouldn't be able to tell this apart from animal flesh."

When people taste new animal-free foods, they often fail to appreciate the significant variation that exists within a single food category like a beef burger. How much and what kind of seasoning was mixed into the meat? Exactly how long was it cooked? What was the fat ratio of the ground beef? Was the cow grass-fed? I would guess that the difference between the Impossible Burger and the average beef burger was similar to that between any two beef burgers the average American eats in a year.

Zahn heavily qualified our tasting with his own view of the product's issues, especially the imperfections that had been fixed in more recent versions. His critiques were precise, highlighting all the specific tastes and aftertastes of the burger, like how long the iron flavor persisted in your mouth. I wouldn't have noticed any of those issues on my own, but surely Impossible's taste-testers—some are vegetarian, but most aren't—have much more refined palates.

The Impossible Burger is built from the ground up using isolated plant fats and proteins to fill the same culinary niche that animal flesh does. It aims to satisfy even the most voracious carnivore. That "meat hunger" has been described by chefs, food scientists, and hunter-gatherer tribes over the millennia. Some cultures differentiate meat and plant hunger, such as the Mekeo tribe of Papua New Guinea, which uses *aiso etsiu*, translating literally to "throat unsweet," to refer to meat hunger, and *ina etsiu*, meaning "abdomen unsweet," to indicate a desire for plant-based food.[24] There's no scientific evidence of a hunger specifically for animal meat driven entirely by biology, but the social forces behind it are very real—from "Beef: It's what's for dinner" ads to the historical association of meat with wealth and prosperity.

Impossible Foods and Hampton Creek represent a huge leap from VegeBurgers and Tofurky. In 2008, Tofurky's founder, Seth Tibbot, explained his product's popularity: "People are happy to have something that is easy to prepare and that can cook right alongside the turkey and is served alongside the turkey. Now everybody's got something to eat. It's kind of a peacemaker product I guess."[25] Contrast that with a 2017 statement by the founder of Impossible Foods: "My company's goal is to wipe out the animal farming industry and take them down."[26]

This bold founder is Patrick "Pat" Brown, a former Stanford University biochemistry professor who left academia and committed himself to solving what he sees as the world's biggest environmental problem. Like many vegans these days, he decided, "it's easier to change people's behavior than to change their minds." Brown felt the food industry was decades behind the curve in biotechnology, leaving wide room for innovation. "The stuff we're doing now that's new to the food system was old news 40 years ago in the biotech world."[27]

His first challenge was to identify the compounds in animal flesh that construct its meaty flavor, so that he could replicate them in plant form. This is no easy task. You can find dozens of active chemicals from hexanal to 4-hydroxy-5-methyl-3(2H)-furanone in beef alone, and these vary widely by the breed of the slaughtered animal, where on their body the meat came from, and even the cooking method.[28] Impossible narrowed its scope by focusing on a specific meat product, Safeway 80/20 ground beef.

The search revealed a key compound that Impossible claims is the holy grail of plant-based beef. It's called "heme," and it's an organic compound with diverse biological function, most well known for its role in hemoglobin, the iron-containing compound that carries oxygen in our blood. Apparently heme is responsible for over 90 percent of beef's flavor, and that gives Impossible a huge advantage in the vegetarian marketplace. Ground beef is around ten parts per million heme, while chicken flesh clocks in at only two. In fact, heme is so beefy that adding it to chicken leads taste testers to think it tastes like beef.[29] On the other hand, a 2011 meta-analysis associated heme in red meat with colon cancer, which adds concerns about Impossible's use of the molecule instead of other plant-based ingredients.[30] The Impossible Burger also has much more saturated fat than other plant-based meats, which could be both a necessary taste factor and a health concern for some people.[31]

Impossible originally found heme in soy root nodules, which are actually colored red from its presence. However, harvesting these and extracting the heme would leave an environmental footprint too big for Brown's taste and would likely come with a prohibitive financial cost. The solution the company found was yeast.[32] If you add heme-coding soy DNA to yeast, the yeast microorganisms dutifully produce the compound for easy harvest. This is the same technology that's been used for decades to produce insulin for diabetics and rennet for cheese.

Brown is also leading a plant-based cheese and dairy company called Kite Hill, but despite his ambition and capability, the animal agriculture industry will probably not be dismantled by a single person, even someone offering a superior product. The animal-free food industry needs collaboration and competition to succeed. Fortunately, Pat Brown has met his match in Ethan Brown, founder and CEO of Beyond Meat, who's been described as "six-foot-five of tanned California muscle with a jawline that could cut diamonds," with the same sort of showmanship we've seen from other industry leaders.[33]

Both companies launched their patties around the same time, but they took different approaches to market penetration. Impossible released its burger only in trendy restaurants such as Momofuku Nishi, a New York Asian and Italian fusion restaurant run by celebrity chef David Chang.[34] Chang is a fitting ambassador given his notorious love of meat. Reportedly when customers complained about his lack of vegetarian options, he responded by adding pork to every dish.[35] Beyond Meat on the other hand went straight to the household, selling packages of two premade burger patties in Whole Foods for around $5.99.[36] (An industry source told me that

Whole Foods and other large retailers wouldn't carry the Impossible Burger because the novel ingredient yeast-made heme lacks the highest standard of FDA approval.) Beyond Meat also has a history of quality plant-based meats, such as its chicken strips released in 2012 that fooled TV hosts in a public, blind taste test. Ethan told me that the final decision to launch the chicken strips was actually made when he and vegan chef Dave Anderson did a casual blind taste test with the company's electrician.

Both Ethan and Pat have found investor and consumer support for their shared ambition of reinventing meat. There is a sense of friendly competition rather than antagonism between the two companies, however. They are both often mistaken for the other, and given how much demand outweighs availability right now, each probably benefits from the other's existence. For outsiders like me who focus on the success of the field as a whole, the rivalry is promising; they are inspired to come up with superior innovations, to work faster, and to build a positive public perception of animal-free meat.

There has been friction around certain issues, usually minor ones such as Beyond Meat adopting a marketing graphic with a cow and a speech bubble that looked similar to one used by Impossible Foods. And competition can have downsides in new industries, such as when rival companies drive down prices to the point where neither is making sufficient profit. We should also worry about what happens if corporate interests overtake moral interests, such as if a big food company buys out either company in order to squash the competition.

But these developments seem unlikely, and the benefits of doing business in vegan tech are far too promising to pass up, so growing competition in the field is welcome. I also see significant value in the different distribution approaches—restaurants versus grocery stores—as both of these have compelling positives and negatives.

And it looks like more competition is indeed on the way, as entrepreneurs are beginning to see the wealth of untapped potential in this space. Other companies are currently growing, such as plant-based meat producer Hungry Planet. The company began when Todd Boyman, a tech entrepreneur and investor, ordered a plant-based burger in a small St. Louis restaurant. When he took a bite, he quickly spit it out and complained to the waiter that he was served someone else's beef burger. The waiter assured him it was plant-based, and as a twenty-eight-year vegan, Todd's entrepreneurial instincts kicked in. He spent months tracking down the origin of the burger. He emailed the producer, forwarding emails about the mainstream news coverage of plant-based meat companies. After about two years of correspondence, in the fall of 2015, he finally met up with the woman behind the burger, Allison Burgess.

Burgess had spent fourteen years working on recipes and selling to local restaurants, focusing on unseasoned plant-based meat products like ground beef and pork so that chefs had as much flexibility with them as they do with animal products. For much of that period, she had worked with vegan chef Freddie Holland, and once the three of them got in touch with Todd's sister Jody, a longtime animal advocate and wildlife photographer, they formed Hungry Planet and began planning out a nationwide expansion, centered on the Midwest. Hungry Planet primarily focuses on food service. One restaurant tried twenty-five different veggie burger recipes before finding culinary relief in Hungry Planet's products.

Hungry Planet hasn't received the press coverage of Hampton Creek, Impossible Foods, or Beyond Meat—indeed, the company prides itself on using minimal technology and making its way without outside investors. Importantly, Hungry Planet shares much of the ambition of the other companies we've talked about: it's focused on replacing animal farming, not on providing premium food options for vegetarians, as plant-based companies have done in the past.[37]

Even more companies are being launched that employ new strategies and technologies like the mass production of plant-based seafood, distributing plant-based meat in India—remember the nation's high vegetarian population—and South Korea, and building a more efficient machine called a Couette cell, which applies the pressure and heat used to make plant-based meat.[38] Plant-based meat is usually made with extrusion, an industrial process by which mixtures of water and plant matter—such as soy and pea protein—are pushed through a long container at high temperatures, after which pressure and cooling align the proteins in the direction the mixture is flowing. These proteins form fibrous structures similar to muscle fibers.

Each of these factors—temperature, pressure, water content, ingredients—can be varied to produce a variety of textures.[39] The Couette cell instead uses a cylinder made of an outer and inner shell. The inner shell rotates while the mixture between the shells is heated and cooled, stretching the proteins into a similar fibrous structure. According to some European researchers, this is more energy-efficient and produces a more meat-like final product than that resulting from the extrusion process.[40]

European companies are also making use of lupin, a legume most well known in the US as a Mediterranean snack. Lupin is similar to soy in many ways, but soy has received significant negative attention due to its ubiquity in standard Western diets and purported health risks. While the scientific community has generally dismissed those concerns as unwarranted—in fact, soy has been shown to have substantial health benefits—lupin has the advantage of avoiding any negative preconceptions.[41] Lupin can also be grown in colder climates than soy, making local production possible in much of Europe. The plants are easier to grow without the use of pesticides, fertilizers, or genetic engine ering; genetically modified organisms have faced much more resistance in Europe than in the US. Lupin also has a higher content of protein, dietary fiber, and antioxidants. There might be some downsides: lupin has a lower yield per acre, and currently, the low production volume of lupin makes it expensive and sometimes hard to acquire.[42] Still, further exploration of its viability seems like a wise move for plant-based meat producers.

The plant-based food industry has now grown large enough to have its own trade group, the Plant Based Foods Association, which was founded in 2016 by Michele Simon, a public health lawyer with over two decades of food policy experience.[43]

At this stage in the development of animal-free food, when it holds only a small fraction of the animal-based market share, it's often quite difficult to know which of these specific approaches is most effective. What's more likely to succeed: the media-focused, tech-focused approach of Impossible Foods, or the minimalism of Hungry Planet? Lupin or soy technology? Or should we be putting our marginal resources into a trade group instead of directly into business and technology?

At this stage, the best approach might just be diversification and innovation. Entrepreneurs or scientists seeing a neglected opportunity should go for it. Throw different strategies at the wall. See what sticks, and then consolidate.

If you can't beat them, join them

The titans of the food industry can be easy to portray as the enemy of innovation and moral progress. It is true that big food companies are the largest purchasers of conventional meat, dairy, and eggs. However, social movements need resources—financial, political, social, and more—to succeed, and these companies have plenty. They have money, factories, expert chefs and food professionals, distribution channels, and other essential tools for moving food from farms to the masses. If these companies start supporting the animal-free food industry, not only does that industry acquire some of animal agriculture's resources, but it might also result in fewer resources going to their other activities—namely, the production of animal-based foods. To put it another way, isn't it harder to take down the multibillion-dollar meat, dairy, and egg industries than to inspire them to switch their production to animal-free versions?

There's already been some investment from food giants, such as the 2016 venture capital investment from General Mills in Kite Hill, the nut-based cheese and yogurt company led by Pat Brown.[44] In 2016, Tyson Foods, the most well-known US meat company, invested an undisclosed amount for a 5 percent share in Beyond Meat. Its CEO later said he thinks plant-based is the future of meat.[45] Beyond Meat's CEO, Ethan Brown, wrote an open letter explaining the decision to let Tyson invest. He noted that Beyond Meat's "most ardent supporters" might feel uncomfortable with the partnership and conceded that Tyson executives' "beliefs don't comport with [his] on animals," but he highlighted the common interest of both businesses in sustainable protein and innovation, as well as the power Tyson brought to the table as a company that "touches 2 of 5 plates in the United States."[46] In fact, even Unilever has hopped on the bandwagon, releasing its own eggless mayonnaise shortly after its arduous and failed opposition to Just Mayo. Its product was cautiously labeled Carefully Crafted Dressing and Sandwich Spread with "vegan" centered on the front label.[47]

An often-overlooked leader in the plant-based food world is Whole Foods and its vegan CEO, John Mackey. Whole Foods was the first retailer to sell Just Mayo, playing a crucial role in Hampton Creek's success, and the first to sell the Beyond Burger, which helped Beyond Meat achieve national media attention and impressive sales. The burgers sold out of the first store to stock them within an hour.[48] Whole Foods has also set precedent for considering the interests of farmed animals with its leading animal welfare standards. The Global Animal Partnership 5-Step rating program is recognized as one of the most useful in the US, though undercover investigations have suggested that the standards and enforcement are still seriously lacking.[49]

It's easy to forget the quiet steps Whole Foods and other retailers have taken by producing their own generic brands of plant-based meat and dairy. While these products usually don't break new ground in terms of media attention or new technology, they provide quality, affordable products for consumers trying to eat more plant-based foods. Even Wegmans, a New England grocery chain with only ninety-five stores, has an extensive line of its own meat-free chicken nuggets, beef strips, and more.[50]

Major investors are also predicting shifts away from conventional animal products. The leader in this field is Jeremy Coller, a longtime vegetarian and major investor in the UK. For his fiftieth birthday, Coller decided to use the influence he had accumulated over his successful career for an ambitious goal: ending factory farming within forty years.[51] Coller launched the Farm Animal Investment Risk & Return (FAIRR) initiative in 2015 "with the aim of raising awareness among investors about the material risks of the massive global growth in factory farming." As of July 2017, FAIRR has $3.8 trillion in assets under management—all committed to the consideration of farmed animal welfare as an environmental, social, and governance issue when the asset-holders decide where to invest.[52]

Similarly, a group of Tyson Foods' investors banded together in 2016 to urge the company to address animal welfare issues in its supply chain and the growing popularity of plant-based foods.[53] Investor pressure might have influenced Tyson's decision to invest in Beyond Meat. These major investors are also following the lead of a small group of trailblazers known as the "vegan mafia," dedicated early-stage funders like Fifty Years and Stray Dog Capital who have been incubating the nascent industry since 2015.

In fact, Tyson is a window into and preview of the broader evolution of the farmed animal movement. The company has been the target of numerous undercover investigations and public pressure campaigns since the 2000s. Yet it was among the first major meat companies to invest in plant-based meat. In a few years, it could be leading the charge for the future of protein, mass-producing the inventions of startups like Beyond Meat. When I ask animal-free food leaders what progress they're most excited about, the most common answer is interest from Big Food. This is the most likely route for animal-free food to take a large portion of the conventional industry's market share.

4. HOW PLANT-BASED WILL TAKE OVER

SMART COMPANIES and impressive products were enough to create today's animal-free food sector, but it will take more to change a significant share of the world's animal consumption. We need to overcome the regulatory and cross-industry entrenchment of animal farming, as well as the biases that keep modern consumers in the habit of eating products of animal exploitation. We also need to weigh different market strategies: Should animal-free food companies stay small and maintain the highest ethical standards, or aim broad and bring in more mainstream investors that could potentially harm the company's mission? Should animal-free food companies focus only on creating products that will reduce the most suffering, such as fish and chicken meat, or products that are easier to make, such as cheese and milk?

To tackle these questions and successfully grow the industry, the movement will need visionary nonprofits and noncommercial actors to coordinate technological development and public dialogue. This kind of advocacy can help companies not just gain traction in the short term, but build a competitive industry that can meet animal agriculture on its own terms. No organization has done more in this regard than the Good Food Institute (GFI).

Wrangling an upstart industry

GFI was the brainchild of Mercy For Animals (MFA), the charity discussed earlier for its leadership in undercover investigations. As an organization committed to the principle of effective altruism, MFA constantly reevaluates its programs and changes course if needed in order to increase its impact. By 2014, it had grown increasingly worried about the ability of social change alone to fix our broken food system.

MFA saw that animal-free food activism gained steam in the US around 2000, but the subsequent drop in animal-product consumption over the next decade was not as significant as many expected—in fact, it really just leveled off the previous steep, upward trend—and there was rapidly increasing consumption in other countries like Brazil, China, and India. If those countries follow the United States' footsteps into the deep piles of factory farm chicken shit, they will endure the same tragedies of animal suffering, resource depletion, and public health crises.

The key strategic insight for MFA was that public opposition to factory farming is far ahead of the public's actual consumption choices, meaning that bridging this gap could be low-hanging fruit for thoughtful advocates. After a year and a half of research by their leadership—particularly Nathan Runkle, Nick Cooney, and Derek Coons—they decided to incubate a new nonprofit organization that supports the development and marketing of animal-free meat, dairy, and eggs. This new organization would focus on business and food technology instead of the traditional activism focus on consumption. While activists had made substantial progress in increasing demand for animal-free foods, the premise of GFI was to put more resources into the relatively neglected strategy of increasing the supply of animal-free foods.

MFA needed to recruit an outstanding executive director to lead the initiative. The nonprofit had its sights set on one veteran animal advocacy nonprofit leader in particular: Bruce Friedrich. At this point, Friedrich had served for thirteen years at People for the Ethical Treatment of Animals as vice president of international grassroots campaigns and for four years as Farm Sanctuary's director of policy and advocacy. Friedrich is currently known in the animal protection community for his cool head and strategic approach. He coauthored *The Animal Activist's Handbook*, and his essay on why people shouldn't support so-called "humane" animal farming was included in *Eating Animals*, one of the most well-known books on vegetarianism.

For decades, he's been known for his willingness to go to great lengths to effect change. In 1993, Friedrich was running a Catholic homeless shelter and a Catholic soup kitchen in Washington, DC. He and three other peace activists, including the famed priests Philip Berrigan and John Dear, decided to take nonviolent action based on the Book of Isaiah's "swords to ploughshares" passage. They trekked through water and rough terrain and snuck past hundreds of Air Force personnel to reach a nuclear-capable Strike Eagle F-15E fighter plane. They proceeded to hammer and pour blood on the plane. On the ground next to the plane, they placed a statement of their intent as peace activists belonging to the Plowshares movement along with a banner that read DISARM AND LIVE.

As one would expect, they were quickly arrested, charged, tried, and found guilty. The jury hearing Friedrich's case took only six minutes to return its verdict. He was sentenced to fifteen months in jail, three years of probation, and $2,700 in restitution.[1]

The furthest Friedrich has gone to help others has actually been his career transition away from these confrontational, provocative tactics and toward the corporate approach of GFI. We might want to answer a moral crime of the magnitude of animal farming with loud, drastic action. That's what it deserves. But that's not necessarily what the animals need. To find out what's best for those we want to help, it is crucial that we apply evidence and reason to find the best way to allocate our limited resources. Friedrich considered this, and he concluded that he needed to go against his gut instincts. He put down his hammer and put on a suit. While some might diverge from Friedrich's approach and decide that the evidence weighs in favor of drastic action and against a more moderate stance, it's important that advocates make evidence-based decisions that utilize both their hearts and their heads.

Since GFI's founding, the nonprofit has built relationships across the animal-free food industry with scientists, investors, entrepreneurs, and other advocates. One of their first projects was mapping out the field with Technological Readiness Assessments, a methodology developed by NASA and now used by various government entities to help organizations understand and evaluate progress in their field. GFI has also helped secure positive media attention for the industry. In the first year, they were featured or had op-eds in the *Washington*

Post, *Wired*, *Vice*, and the *Los Angeles Times*.[2] In early 2016, GFI helped launch Memphis Meats, a startup developing cultured meat—real meat made without killing animals—earning that company coverage in the *Wall Street Journal* and other major media outlets.[3]

While MFA was launching GFI, it was also creating a second organization, the $25 million venture capital fund New Crop Capital (NCC). NCC has funded Memphis Meats and other companies supported by GFI, focusing on early-stage investments for companies working to disrupt the animal agriculture industry. The one-two punch of these organizations provides a near-ideal environment for animal-free food companies to incubate and flourish.

Which products do we prioritize?

One question the movement faces is which products to focus on first: beef, chicken meat, dairy, eggs, or a nonfood product like leather. While this might seem mundane, there is actually huge variation in the amount of good one can do by working on each product.

First, we need to evaluate the harm caused by each product by putting together (1) the calories per animal (you get a lot more meat from a cow than a chicken); (2) the lifespan of each animal on the farm; and (3) an economic factor called "elasticity," which is the estimate of how much a change in one unit of demand affects production. Putting these together, we get the following: per five hundred calories consumed, chicken meat leads to an estimated average of 3.0 days of those animals living in factory farms or similar conditions; turkey causes 0.8 days; battery cage eggs cause 6.3; and farmed fish causes approximately 27.5 to 159.1, varying based on species and location. Meanwhile, perhaps surprisingly, beef and pork consumption cause 3.8 hours and 7.6 hours, respectively, and five hundred calories of milk consumption is estimated to cause about 17.5 minutes.[4] The trend here is that consuming smaller animals leads to far more suffering per calorie because it takes far more animals.

We also want to consider the varying treatment of each animal on the farm, during transport, and at slaughter, as well as environmental and human health considerations. The calculations for those are beyond the scope of this book, but the result seems to be that the animal products maintain a similar ranking of harm caused, with the size of the animal still being the dominant factor.[5]

One exception to the size rule is the unusual case of bivalves—particularly oysters and mussels. These two creatures are animals by definition, yet they lack most of the behavior and neuroanatomy that we typically associate with consciousness. Instead, they are more like plants, sitting in one place and reacting in relatively simple ways to their environment. For this reason, some thoughtful eaters identify as "bivalvegans," vegan with the exception of eating bivalves—usually just oysters and mussels because those species tend to be immobile and therefore have less likelihood of sentience.

In addition to the harm caused, we need to consider social, technological, and culinary factors. For example, we know that dairy—in particular, cheese—is so often cited as the deal-breaker for would-be vegans. This means that high-quality plant-based dairy could be a powerful social lever. Additionally, some dairy and egg products like mayonnaise are much easier to produce with plants because they have a homogeneous texture and significant nonanimal-derived flavors, such as lemon juice. So starting with them could more easily establish the viability of plant-based foods. Consumers might also be more open to swapping out animal-based foods when the animal-based ingredients are less salient, as well as in nonfood situations like wearing shoes made from plant-based leather.

All things considered, it seems the movement should prioritize reducing the consumption of chicken meat, eggs, and fish meat because of the massive harm associated with these products. However, in the next few years as plant-based foods are still establishing a consumer foothold, it makes sense to focus on dairy and egg products that are much easier from a culinary and technological perspective.

"Follow Your Heart" or "follow the impact"

In 1970, four young vegetarian men with long hair and bushy beards were each hoping to find purpose in Southern California: Bob Goldberg, Paul Lewin, Michael Besançon, and Spencer Windbiel. Besançon had just traveled to Mount Shasta in Northern California because it was "believed to be a focal point for spiritual energy."[6] Inspired by this journey and a natural food conference, he decided to start a vegetarian lunch counter in the back of a health food store founded in Los Angeles by Johnny Weissmuller, the actor and Olympian swimmer who portrayed Tarzan in movies of the 1930s and 1940s.[7]

In 1973, the store was up for sale and the four men bought it with the goal of taking all meat products off the shelves. They renamed the store Follow Your Heart. Despite naysayers, success came quickly. In Goldberg's words: "Instead of crashing and burning immediately the store became very successful. I think what people hadn't realized was how comfortable it is for somebody who doesn't want to be around those [meat] products to go to a store where they can eat anything."[8]

When the store's supplier of eggless mayonnaise went out of business, Goldberg went to work on a replacement. That mayo went on to become Vegenaise, a longtime vegan favorite and Follow Your Heart's flagship product, and the team started developing other products. Eventually, the company shifted its main focus to product development.

Today its product lineup includes block cheese, shredded cheese, cream cheese, ranch dressing, blue cheese dressing, and other plant-based dairy products. The company targets health-focused consumers who want non-GMO, organic, gluten-free, or soy-free foods. Their production facility is called Earth Island and is powered by an array of rooftop solar panels.[9] Goldberg told a reporter in 2016 that his inspiration to replace eggs with plant protein came to him in a dream.[10]

Follow Your Heart continues to innovate in important ways. In 2015, it released a new product, VeganEgg, which is made from algae and is probably the best animal-free egg product on the market today, often used for scrambles and omelets. Despite Follow Your Heart's success, Goldberg has rejected every offer from outside investors wanting to cash in on the growing demand for animal-free foods. He says, "It's very difficult for companies to grow and maintain their values. And I'm very focused on that."[11]

When I spoke to Goldberg on the phone, his commitment to purity and high-quality food was palpable. He was clearly dedicated to making Follow Your Heart a wholesome, immaculate company, and many vegans over the past few decades have been thankful for that. But thinking about the total impact the company could have in reducing the harms of animal farming makes me worry that his focus on purity might come at a significant cost to our food system. Even if a focus on growth were to lead to some sacrifice in values—and I'm honestly not sure it would—I suspect that the impact from the acquisition of new customers and the expanded reach of the company's ideology of compassion would quickly outweigh any loss from the perspective of the desperate animals in our food system. While value-centric companies like Follow Your Heart have been the saving grace of dairy-hungry, high-income vegans over the past few decades, their approach might come with high opportunity cost. Those forlorn animals don't care about our purity. They just want to be free from suffering by whatever means necessary.

Some advocates worry that giving any ground on their end goal, such as by using nonorganic ingredients if they are aiming for an organic food system, could make it harder to achieve that ideal outcome. There is a case to be made for radicalism, as it could push public discussion forward, but it seems to me that there's a best-of-both-worlds solution for most advocates to express radical goals while supporting practical, sometimes partial solutions.

Another company thinking about artisan appeal and big-picture change is Miyoko's Kitchen. Miyoko Schinner, the founder and chairman of the company, didn't eat much dairy as a child living in Japan. One day after she moved to America at the age of eight, she got to try pizza at a friend's party. In her words, she was excited to become "a true American." At first, she hated the gooey, fatty cheese, but like many Americans, she quickly became hooked. The fried rice and bento boxes she brought to lunch were oddities in California's schoolyards, and her mother took cooking lessons from a neighbor to learn to cook American fare for Schinner's father.

Schinner went vegetarian for ethical reasons, and eventually transitioned to veganism for health reasons while living in Japan again after college. She had developed a passion for French and Italian cuisine at this stage in her life, so she decided to apply the lessons she learned from those areas to plant-based baking and, eventually, cheese-making after a long series of entrepreneurial successes and failures in the animal-free food industry.[12] Now she uses bacteria cultures to ferment nuts, to break their sugars down into acids in the same way bacteria are used in animal-based cheese production. That's the main difference between plant-based and animal-based cheese production: whether the fats, proteins, and carbohydrates you begin the process with are from an animal's milk or a plant ingredient like a nut. The choice of bacteria, the duration of the aging, and additions of vegetable starches or fats allow her to produce a wide variety of cheeses, many closely emulating well-known cheeses like mozzarella and Brie.

As of February 2017, Schinner was producing one hundred thousand rounds of cheese a month and was planning to release versions of the company's cheese that would match the average animal-based cheese more closely in both price and nutritional content.[13] However, Schinner sees her company as kind of like Tesla. Tesla first came out with its premium roadster several years ago and is only recently lowering its price point. Miyoko's Kitchen will eventually produce cheaper block cheeses than the artisanal rounds it makes today, but even these will always be higher-end foods. She told me they'll never make cheese shreds or slices, and the ingredients will always be organically sourced.

Schinner and I agree that in order to effect big changes in our food system, we will need a diversity of companies and approaches. Just like it is in the existing dairy industry, there will be high-end and medium- to low-end products. But aspiring entrepreneurs, investors, and researchers may want to note that we already have Miyoko's Kitchen, Follow Your Heart, and many other companies producing exceptional high-end products. So despite the significant impact of those artisan companies, what we need most, I think, is people tackling the mainstream animal products—those that make up most of our consumption—with plant-based alternatives.

Labeling, marketing, and shelf space

One roadblock that is probably slowing down mainstream acceptance of plant-based products, even artisan ones, is labeling. When the California Department of Public Health inspected Schinner's production facility, the agent saw that the product was labeled only according to flavor, such as Aged English Fresh Farmhouse. It couldn't be categorized as cheese, so the agent asked her for the actual name of the product. Schinner, on the spot, decided to call it a cultured nut product.

While this was a snap decision, it has stuck, though Schinner is now moving toward mainstream dairy titles for her products when possible. For example, when her company launched its first butter in 2016, the name was unabashedly European Style Cultured Vegan Butter. This name does include the word "vegan," but that more reflects Schinner's desire to "hold the vegan banner high" than reservation about using the term "butter" alone.

Plant-based milk producers are already facing legal challenges. The standard of identity for milk defines it as "the lacteal secretion, practically free from colostrum, obtained by the complete milking of one or more healthy cows, which may be clarified and may be adjusted by separating part of the fat therefrom; concentrated milk, reconstituted milk, and dry whole milk. Water, in a sufficient quantity to reconstitute concentrated and dry forms, may be added."[14]

These laws are necessary to help consumers easily identify different food products. If there were no criteria defining what makes ketchup, for example, we would have to constantly watch out for companies trying to peddle ketchup products made with improper ingredients or production methods.

In the preindustrialized world, it was much harder to fake food. Producers had little opportunity to substitute lesser ingredients, and most consumers grew their own food or knew the local farmers. They could often inspect the bakery, the crop field, or even the slaughterhouse if they wanted. Even then, food was subject to regulation. In ancient Rome, bread was sold on steps, and the higher the step, the higher the quality and price.[15] This was originally done informally by merchants, but became required in the Theodosian Code of 438 CE. If you allowed bread to be moved from one step to another, you risked "the severest punishment."[16]

This continues today. In 2011, two Spanish businessmen were sentenced to prison for selling "olive oil" that was actually 70–80 percent sunflower oil.[17] In 2008, Chinese fraudsters added water to cow's milk used to produce infant formula while using the chemical melamine to increase the apparent protein content of the formula when tested. Nearly three hundred thousand babies fell ill, approximately fifty-three thousand were hospitalized, and six died. Eleven countries stopped dairy imports from China after the incident.[18]

Such profit-driven food crimes are good reasons to enforce strict labeling standards. But the animal agriculture industry has also tried to twist the intent of these laws to drive up their own profits. Take "soy milk," for example. This is an established product name that identifies a white, milky beverage made from soy. I've never heard anyone wonder whether "soy milk" refers to soy-flavored cow's milk. Nonetheless, the dairy industry is waging a campaign to prevent plant-based milk producers from using "milk" to describe products such as soy and almond milk. They want the standard of identity enforced, seemingly just because it would harm their upstart competitor. It also seems the cow's milk industry is willing to throw other animal products under the bus: because the definition of milk specifies that it must come from a cow, the implication is that the beverage that comes from goat udders needs to be called something like goat juice.

These labeling efforts are supported by US congressional representatives from high dairy production states. As of this writing, they have received little support from other legislators, but this serves as a reminder of how industry affects policy and should make us hopeful about the support we can obtain for animal-free meat, dairy, and eggs once they become significant parts of at least some state economies.[19]

The fact is that the public's perception of "milk" is no longer aligned with its outdated legal definition, and food standards should be updated to reflect that. When the media reported on the 2016 and 2017 efforts by US congresspersons from agricultural states to enforce the strict definition of the term, they spent less time in their articles discussing the proposed rule change than they did on the growing popularity of plant-based milks. An article by the *Los Angeles Times* editorial board used the headline "Got 'Milk'? Dairy Farmers Rage Against Imitators but Consumers Know What They Want," and *Yahoo! Finance* reported, "Dairy Farmers Are Losing the Battle over 'Milk.'"[20]

In 2015, there was 9 percent growth in plant-based milk compared to a 7 percent decrease in dairy milk sales, making the plant-based milk market 10 percent the size of conventional milk. The dairy industry feels threatened, and it's lashing out by any means possible.[21]

Plant-based milk has been the most successful animal-free product on the market, despite the fact that it's one of the least emulative products in terms of nutrition. Soy is the only popular one with protein similar to animal-based milk. Plant-based milks often have lower sugar content than milk from animals, but the sweetened plant-based varieties—which have approximately the same sugar content as animal-based—are more popular. Some worry that added nutrients in plant-based milk like calcium are less easily absorbed by the human body, but others think this is misguided and that absorption rates are similar in both products.[22]

In the meat market, the leading plant-based products have nutritional and culinary profiles quite similar to animal-based products. Ethan Brown, CEO of Beyond Meat, argues that it makes sense to call his products "meat." In an interview with TV personality Dr. Oz, he explained: "We like to use the language of plant-based meat, and what we're doing is, we're taking all of the core constituent parts of meat. We're taking those directly from plants: basically protein, fat, and water. We're assembling those in the architecture of meat or muscle, and we're providing it to consumers in that form. So they're getting a piece of meat in terms of its constituent parts. It just doesn't come from an animal."

The only differences, Brown argues, will be the nutritional benefits of the Beyond Burger. For example, it lacks cholesterol, which despite being a common feature of animal meat, has no noticeable impact on taste or texture.[23]

There's also historical and contemporary precedent for using terms like "meat" outside of animal products, such as coconut meat, nut meat, and even the "meat of the matter" to refer to the substance of an issue. We also say "peanut butter" and "cocoa butter," terms that certainly aren't confusing consumers. To refer to plant-based meats as "fake" or "alternative" is not more accurate; it implies that animal-based meats are the gold standard in a way that doesn't properly reflect the ethical, health, and taste considerations, and doesn't reflect the commonsense use of the relevant terms. In fact, a few years or decades down the road, we can hope to see labels and terminology that help consumers understand the harms of animal-based foods, similar to the cautionary text on cigarette cartons. We've already seen some restrictions on the misleading positive labels like "humanely raised," though this is usually done without an explicit label, such as with picturesque farm images that in no way actually reflect the appearance of the vast majority of modern farms.

Overall, I think there's a good case for calling the Beyond Burger "meat" without qualification. However, it would be concerning to me at this time if companies called soy milk, almond milk, and especially a product with a significantly different nutritional profile like coconut milk—tasty as it is—simply "milk" without identifying the plant it's derived from. But remember, that's not important for the current debate over the term, which concerns products such as those that are labeled "almond milk" and have pictures of almonds on the packaging. In those cases, it seems clear that consumers know what they're buying and the dairy industry is simply trying to hassle a competing industry in an effort to bolster its tumbling sales. If the dairy industry has a genuine concern that consumers are missing out on protein, I'd note that the average American consumes far more than the Recommended Daily Allowance of protein, around 145 percent the RDA for women and 176 percent for men.[24]

We should also consider that our terminology is a way for producers to convey important social information about their product. By using "meat" to refer to plant-based foods with the same taste, texture, mouthfeel, and nutritional profile as animal flesh, we are telling people that they can get all those features without the animal cruelty, environmental devastation, and negative health impacts.

At the time of this writing, Beyond Meat labels its burgers as plant-based without using the term "meat." Ethan Brown told me that the company's current focus is on perfecting the product, and once it does that and public opinion data shows consumers are on board with this use of the term, they might switch. This seems like the right call because consumers are still getting familiar with plant-based meats, and regulatory issues at this stage would be a big hassle. As the industry establishes itself and public attitudes shift, updated labeling will be an important stepping-stone on the path toward an animal-free food system.

Utilizing the obsession with animal-based foods in order to promote animal-free foods can go too far. SuperMeat, a cultured meat company I will discuss in the next chapter, has used the line "Meat is delicious! So delicious, in fact, that we're not going to stop eating it."[25] I think this is bad for the movement. Messages need to be kept simple, and most people who see this one probably won't appreciate that the "meat" it's referring to actually isn't made from animals. Even if they do, the message still implies that giving up meat is a loss. Fortunately, this sort of advertising is rare.

Another example of animal-free advertising that might do more harm than good is a campaign by Ripple Foods. Ripple is a pioneering milk company that utilizes a patented protein-purification process to extract plant protein without bringing along the plant flavor. They use yellow peas for this, the same vegetable Hampton Creek uses for its mayo. Ripple refers to its product as "dairy-free milk" instead of identifying its specific protein source on the front of its package. This is easier to do for peas than nut-derived milks because peas are not a common allergen, but mainly it serves to emphasize that the product is not a plant-flavored product and to allow flexibility for Ripple to select different on-trend ingredients in their manufacturing.[26]

In general, Ripple is a very exciting company, especially given that its founders both helped lead and learned lessons from two other cleantech movements: Neil Renninger with renewable fuels and Adam Lowry with sustainable household-cleaning products. When I spoke with Renninger in his company's lab near San Francisco, he seemed emboldened by the failures of renewable fuels, eager to avoid that movement's mistakes. By and large, I think Ripple is doing that successfully. Renninger was particularly excited to move into food technology because the food and beverage industry spends an extremely low amount on R&D, around 1 percent of sales, while other sectors spend closer to 5–10 percent and many innovative Silicon Valley companies spend up to 20–25 percent. It's an industry ripe for innovation. When I tried Ripple's milk and yogurt, the products definitely stood out to me as some of the closest replications of animal-based milk on the market today, though a fellow taste tester disliked the milk, remarking that it tasted like peas.

My concern about Ripple is that its marketing strategy includes disparaging some of its plant-based colleagues. In response to the dairy campaign to restrict the use of the term "milk," Ripple launched a website saying:

> DEAR DAIRY, I can understand why you're upset. Almond milk is a sham. Only 1g of protein and less than a handful of almonds in an entire bottle? That's not milk. Cashew and coconut milk are even worse; they don't have any protein.

The campaign shows viewers an 8-bit-style game where the user answers questions about what milk "should be," such as, "Should milk shower you in sugar?" It displays an ongoing race next to the questions, and with each answer you give, Ripple inches farther ahead of the competition. Oddly, even if you say you want the sugar and other seemingly undesirable features, it still concludes that Ripple is a better fit.[27]

I think this is a poor approach. While it's true that Ripple is a great plant-based option for consumers who want the same nutritional profile as dairy without dairy's downsides, I think the animal-free food industry thrives and withers together. If consumers are genuinely misinformed about the nutritional aspect—I haven't seen data that they are—then educating them is useful, but not through attacking those products with an adversarial marketing campaign. This is especially true if the misinformation is about protein, as Renninger has suggested, given the typical overconsumption of protein in the American diet. Negative campaigns like Ripple's might boost short-term attention of an individual company and perhaps profits as well, but runs a significant risk of impeding the whole industry's long-term growth—and likely even the growth of the company utilizing the negative messaging.

Pat Brown, CEO of Impossible Foods, has made a similar mistake in my opinion in calling cultured meat "one of the stupidest ideas ever expressed."[28] He's not an expert on this technology, and clearly many of his scientific peers are much more optimistic. In the field of cellular agriculture—growing agricultural products with cells rather than whole organisms—some of the entrepreneurs I interviewed told me about issues they had getting funding in Silicon

Valley explicitly due to Pat Brown's negative views and influence in that community. Pat's other comment on cultured meat, that "you buy into, at best, the same limitations that a cow has," makes it clear he doesn't really understand the technology, which is exciting precisely because it frees us from the limitations of animal biology. I'm glad Pat Brown is working hard on plant-based meat, but his sweeping dismissal of promising alternative technology is unwarranted.

An issue similar to labeling and advertising, and one that we likely must tackle for the industry to successfully jolt consumers out of the status quo, is shelf space and placement—literally where the products sit in the grocery store. This was key in the stunning sales growth we've seen of plant-based milks. In 2002, Dean Foods, a massive food and beverage company specializing in dairy products, bought White Wave, makers of the popular Silk soy milk brand.[29] With the resources of Dean Foods, Silk got additional shelf space in grocery stores—specifically, space right next to the animal-based milk. This simple change made the product much more accessible, and as much as we consumers want to believe that we are not influenced by something as trivial as shelf placement, having to walk across the grocery store to find a new product is a big deterrent.[30] Beyond Meat has been the first plant-based meat producer to achieve such premium shelf space, now stocking its patties in the meat case at Whole Foods.[31]

There is also psychological evidence that vegetarian food will benefit from being incorporated into restaurant menus' main selections, as opposed to being listed in a separate vegetarian section. Changes like this should also be a priority of animal-free food companies and activists.[32]

Plant butchers

In addition to the large plant-based companies discussed thus far, small businesses are taking their own piece of the pie. Take, for example, the vegan sister-brother duo of Aubry and Kale Walch—yes, that's his real name. One day in casual conversation with friends, Aubry half-jokingly suggested that they open a vegan butcher shop. The idea grew on them, so Aubry and Kale—just thirty-four and twenty-two, respectively, at the time—worked together to launch the Herbivorous Butcher at the Minneapolis Farmers Market in June 2014.

The butchers quickly earned a loyal and dedicated local fan base thanks to their quality cuts like Italian sausage and Huli Huli Hawaiian ribs, as well as their hook of being America's first vegan butcher shop. The duo's chipper personalities helped too. This momentum enabled them to raise over $60,000 on the crowdfunding website Kickstarter to open a brick-and-mortar storefront.[33]

Since their Kickstarter campaign, Aubry and Kale have received international media coverage in top outlets including NPR, *Time*, the *Telegraph*, and the *Guardian*. They were even featured in an episode of *Diners, Drive-Ins, and Dives* with celebrity chef Guy Fieri.[34]

While the Herbivorous Butcher might not grow to become the next Beyond Meat, their team is impact-driven and has made a significant difference in their local community, as well as to their online customers and the audience they reach through media coverage. As the animal-free food economy grows, there will be room for an army of small businesses like theirs, who could collectively form a branch of the movement every bit as strong as the bigger companies. So motivated readers should take note—what's your business idea?

5. THE WORLD'S FIRST CULTURED HAMBURGER

ON FRIDAY, AUGUST 2, 2013, the temperature in the Dutch city of Maastricht reached a record high of 94 degrees.[1] That afternoon, Mark Post biked home from the tissue-engineering laboratory at Maastricht University with around $600,000 worth of meat in an icebox on the back of his bike. He was carrying two of the first hamburgers ever made from cell-cultured beef: meat grown from animal cells outside of an animal's body. They were to be shipped to London over the weekend for a televised public tasting on Monday.[2] To many advocates, this event felt like the beginning of the end of animal farming.

The technological ability to grow meat like this might be new, but visionaries have been expecting this development for at least eighty years. In his 1931 essay "Fifty Years Hence," Winston Churchill predicted: "We shall escape the absurdity of growing a whole chicken in order to eat the breast or wing, by growing these parts separately under a suitable medium." The scientific foundations of this prediction began decades prior, in 1885, when zoologist Wilhelm Roux extracted tissue from a live chicken embryo and maintained it in warm saline solution for several days.[3] Starting in 1912, the French surgeon Alexis Carrel sustained a piece of embryonic chicken heart tissue for thirty-four years.[4]

There was little progress for the next few decades, which didn't bode well for Churchill's prediction. That changed in the 1970s, when researchers started growing muscle fibers in vitro, meaning they extracted small samples of tissue from animals, placed them in culture media—a bath of nutrients and growth molecules for the cell—and saw a significant size increase. Some scientists such as

Russell Ross did this with aortic tissue from young guinea pigs to better understand atherosclerosis, a disease characterized by plaques in the arteries which can lead to heart attack and stroke.[5]

"Cultured meat" is the term I'll use for any animal cells grown outside an animal's body for human consumption, but it's sometimes referred to as cell-cultured meat, cell-based meat, in vitro meat, lab-grown meat, and clean meat. By 1981, the year Churchill predicted that we would be growing cultured meat, no scientific team was working to culture muscle fibers for human consumption. It wasn't until 1995 that another advocate, Willem van Eelen, took up the cause by filing patents for cultured meat in the US, the Netherlands, and other countries. Van Eelen was held as a prisoner of war as a seventeen-year-old soldier in World War II, and during that period he became intimately familiar with intense human suffering—especially hunger, as he was tasked with divvying out rice to his fellow prisoners. He witnessed his captors abuse animals in much the same way they abused him, sparking a lifelong interest in reducing animal suffering.

After his release, Van Eelen went to university in the Netherlands and met a biology professor who kept alive a large piece of meat in vitro, like Carrel had done in 1912. After college, Van Eelen managed art galleries and restaurants, but he toyed with the idea of cultured meat for most of his adult life and attempted some unsuccessful lab work that led to patents in the late 1990s.[6]

The next milestone in the field came in 1998 when a team of American engineers funded by NASA grew goldfish meat in vitro. Their goal was to better understand the viability of cultured meat as a way to feed astronauts. The fish fillets they successfully produced were fried with olive oil, garlic, lemon, and pepper, but the researchers never tasted them since consumption wasn't allowed by the FDA, presumably because it hadn't been tested for food safety. Lead engineer Morris Benjaminson observed that "they looked and smelled just like fish fillets." The NASA research was discontinued because the cell-based food was expensive to produce and seemed like too much of a moon shot.[7]

The first cultured meat that people admit to eating was part of the work of Australian artist Oron Catts in 2003. In 2000, he prototyped a "semi-living" steak from prenatal sheep skeletal cells, and three years later his team exhibited at L'Art Biotech in France with cultured frog meat, serving it to six guests to draw attention to the "kind of hypocrisy [needed] in order to be able to love and respect living things as well as to eat them."[8] This work and similar art projects are drawing praise from some in the cellular agriculture field for exploring the boundaries of this fascinating new technology.

Another artistic team produced *The In Vitro Meat Cookbook*, which includes illustrations of "knitted meat," meat ice cream, and other strange cultured-meat dishes.[9] I worry that this sort of art makes the technology seem a bit absurd, more like a quirky experiment than a viable solution to our broken food system. That being said, it is fun to imagine the possibilities.

In 2004, Van Eelen—who referred to himself as "the godfather of in vitro meat"[10]—kicked off the modern era of cultured meat research. The Promethean businessman contacted scientists at Utrecht University and Eindhoven University of Technology in the Netherlands, trying to convince them to apply for a research grant from the Dutch government. The scientists were persuaded by the "very charismatic" eighty-one-year-old entrepreneur, as was the Dutch government.[11] He also recruited Peter Verstrate, a meat industry professional, to serve as the research program administrator.

As the research project began, the aforementioned Mark Post was working one day a week at Eindhoven on unrelated research, but when the local cultured meat program manager fell ill, Post was happy to step in. He supervised two PhD students and found that most of the other scientists were working on the project for reasons other than food production, such as medical applications of cell-culturing technology.

Eventually, the press caught wind of the research and began contacting the scientists for quotes and information. One reporter from the *Sunday Times*, the largest traditional newspaper in the UK, tried contacting two scientists in 2009 before finally reaching Post. That happenstance coverage made Post, a relative newcomer, the go-to contact for the topic in the flood of news articles that followed.

During this deluge, Post received an email from Rob Fetherstonhaugh, whom Post assumed was just another reporter. Fetherstonhaugh asked the usual questions about the research, like where the cells came from, and asked if he could visit him in person. Again, not an unusual request. On May 5, 2011, Post met with Fetherstonhaugh and discussed his ambitious plans for growing the field. Post wanted to raise interest in cultured meat with a public tasting, perhaps of a few links of sausage—a relatively easy meat product compared to steak or pork belly. He envisioned the taste testers on stage with the happy and healthy pig from whom the cells were taken. Post hoped such an event would generate enough funding to scale up and commercialize the technology.

At the end of the conversation, Post was happily surprised to learn that his visitor represented the cofounder of Google, Sergey Brin, who wanted to fund the public tasting as part of his mission to support world-changing tech ventures.

Eventually, Post and Brin's team settled on a hamburger prototype and decided not to bring the cow onstage.[12] The timeline for the event grew longer due to unexpected issues like the meat not binding together properly at the scale of the burger—monumental compared to the tiny scientific samples previously made in the lab. Lab technicians combined muscle fibers from thousands of these samples, totaling around forty billion cells, with red beet juice, saffron, bread crumbs, and egg to create the hamburger patties. This process is very labor intensive. Indeed, some of the scientists I've spoken with in the field find the process of creating prototypes to be quite frustrating. It takes a huge amount amount of work and, while it's useful for funding and publicity, it rarely helps generate any real scientific knowledge.

The media used to grow the fibers included fetal bovine serum, an ethically problematic and expensive product made from the blood of cow fetuses. While it's used in research, this component is not viable for commercial production as it's prohibitively expensive. Researchers have instead focused on inexpensive plant-based versions, though improving them is an active area of research.

Finally, on August 5, 2013, the tasting setup was ready. The hamburgers stayed intact, even while being transported by bike on the hottest day of the year.[13] The public tasting opened with a video of Brin, an anthropologist, and an environmentalist hyping the transformative nature of the technology. Post, the chef, and two journalist taste testers shared their excitement about the burger, which was carefully cooked with sunflower oil and butter. Then the tasting began. The first reaction, after a long chew, was, "I was expecting the texture to be more soft," followed by concerns about the lack of fat and seasoning. But the overall conclusion was that it was "close to meat" and that it was certainly impressive as a first step, especially given the burger itself was only muscle cells—no fat.[14]

New Harvest

The Good Food Institute has helped grow the plant-based and cellular agriculture industries since 2015, but the nonprofit organization New Harvest has been hand-rearing cellular agriculture since 2004. If Van Eelen is that industry's godfather, New Harvest's founder, Jason Matheny, is the doctor who delivered the baby and provided expert care in its infancy.

Prior to 2004, Matheny earned his master of public health from Johns Hopkins University and worked as a consultant for the World Bank and Center for Global Development. Like many people who take an interest in effective altruism, he was motivated by the utilitarian philosophy he was exposed to in university. After Matheny visited a poultry farm in India, he became convinced that he needed to tackle the issue of animal farming. He read about the NASA goldfish experiments and contacted all sixty of the authors cited therein to discuss the possibility of large-scale cultured meat production.

Matheny realized that cultured meat sat in an academic no-man's-land at the intersection of food production and medical technology, the latter of which is where most tissue-engineering research occurs. This neglect made it difficult for researchers to find sustainable funding and institutional support, so Matheny founded New Harvest in 2004 to help.

New Harvest's first big win was helping Van Eelen convince the Dutch agriculture minister to support the research effort he facilitated. In 2006, New Harvest raised an additional sum from private donors to support the research, and in 2008, they helped with the world's first International In-Vitro Meat Symposium in Norway. New Harvest continued hosting workshops and coordinating research all the way until the cultured hamburger launch in 2013. In fact, they were the ones who originally referred Sergey Brin to Mark Post as a potential grantee. At that point, Matheny moved on to a new high-impact role in intelligence research for the US government, based on the same effective altruism motivations, and another young leader took the wheel at New Harvest.

In 2009, Isha Datar was in her senior year at the University of Alberta, where she was studying cellular and molecular biology. She took a graduate-level course on meat science and was galvanized by what she learned about the harm of animal farming. After her professor attended one of New Harvest's seminars and told the class about the prospect of cultured meat, Datar wrote her term paper on the topic, which snowballed her enthusiasm. She reached out to Matheny, who helped her publish the term paper in the journal *Innovative Food Science and Emerging Technologies*.

Since Datar took over as executive director in 2013, New Harvest's work has kicked into high gear. One of her first projects was founding a company to work on an easier project than cultured meat: cultured milk. She reached out to two people in her network who had expressed an interest in making real milk without cows, recent graduates Ryan Pandya from Tufts University and Perumal Gandhi from Stony Brook University. These three young scientists had only four days to coordinate before the application deadline of a biotechnology-focused startup accelerator that could provide them with lab space, mentorship, and $30,000 in seed funding.

They were accepted as the New Harvest Dairy Project and quickly packed their bags to move to Ireland for the summer. Thanks to the hype around Post's burger and New Harvest's workshops, biotech investors were eager to hop on board. Then journalists learned of the project and a media barrage followed, which led to more investor interest. The team then traveled to Hong Kong to meet with Horizons Ventures—the venture capital firm funded by Li Ka-shing, whose investment in Hampton Creek was discussed in an earlier chapter. In one summer, the company, now called Perfect Day, received more attention than many successful startups do after several years of existence.

Perfect Day—originally launched as Muufri after its incubation period—claims the process for milk is much easier than that for cultured meat, primarily because milk is acellular. This means that instead of growing whole cells, Pandya and Gandhi only have to grow the dairy proteins casein and whey. You might know casein as the culprit of cheese's addictive nature, and whey for its popularity in dairy-based protein shakes. Perfect Day makes these proteins with the same process used to make insulin for diabetics and to make chymosin, the main constituent of rennet, a common cheese ingredient originally taken from the stomachs of young calves. It's also the same process used for the heme in the Impossible Burger. The process involves taking the genes that encode a molecule's production—whether that's casein, whey, insulin, or chymosin—and inserting them into microorganisms like yeast. The microorganisms are not included in the final product, so the final product is the exact same molecule you would get from farming cows for their milk, refuting any health concerns. For insulin, this gene-insertion process has made the life-saving medicine much less expensive. For animal products, it provides a useful tool for fixing our broken food system.

Before founding Perfect Day, Pandya also toyed with the idea of making an artificial udder with tissue-engineering technology, but the yeast idea was much simpler with more commercial precedent. They even named their dairy-yeast Buttercup.[15]

In 2016, I had the opportunity to visit Perfect Day's headquarters in Berkeley. The company had just moved into a new location, where I had to slide past workers giving the entrance a fresh coat of paint. It was a typical open-office Silicon Valley startup headquarters, but in the back was a laboratory space that looked just like the lab desks in a high school chemistry classroom.

Pandya told me how in high school he learned about the immense suffering of farmed animals. The connection between his dinner plate and animal cruelty made him deeply uncomfortable, and he considered vegetarianism, but he also faced peer pressure that ultimately dissuaded him. Then Pandya spent the summer before college as a camp counselor with a friend who was also interested in vegetarianism. Since college was a fresh start, Pandya and his friend took the plunge.

That fall, he read *Eating Animals* by Jonathan Safran Foer and realized that the animal cruelty arguments for vegetarianism also applied to replacing dairy and eggs, so he became "vegan-ish." His compromise was that one day a week he could enjoy an eggplant Parmesan, but after each meal, he'd question if this small indulgence was worth it. Later in his college career, he took a course on tissue engineering and had his "lightbulb moment" when he considered the possibility of engineering animal-free animal products.[16] He saw this as a high-impact role in the movement that might be neglected because it requires not just ethical motivation but also scientific knowledge and an entrepreneurial spirit—a trifecta that describes Pandya perfectly.

Pandya contacted New Harvest for advice, and after graduation he worked on acquiring lab experience by working for a small biotech firm that made vaccines using the same gene-insertion process that Perfect Day would end up using. He moved into an entrepreneur house with the pitch of making an animal-free milk company. That's when Datar told him about the accelerator deadline and connected him with his cofounder Gandhi, another young scientist who went vegetarian in sixth grade after receiving a provegetarian leaflet in India.

A laboratory fridge at Perfect Day's office held a dozen or so samples of yogurt in small mason jars. As with the Beyond Meat and Impossible Foods burgers, I struggled to find something to complain about when I tried one. In my opinion, plant-based yogurts already mimic animal-based yogurt pretty effectively. Perhaps the one distinction I could notice was that this yogurt had more of the fat and richness common in the average dairy yogurt, similar to what whole milk has that skim milk lacks. I think most plant-based yogurts have particularly low fat content because the companies want to appeal to health-focused consumers. Like my friend at Impossible, Pandya was far more critical of his own product. He estimated that Perfect Day was 70 percent of the way to matching artisanal animal-based yogurts. They had figured out most of the tongue flavor like sour and salty notes, but weren't quite there with the retronasal depth of milk fat, and they hadn't fully captured the subtle grassy notes of their target brand, Liberté. Fortunately, their Head of Food Development, Ravi Jhala, has fifteen years of experience working in the flavor department of big dairy brands like Sargento, Blue Bunny, and Chobani, so I'm optimistic that they'll reach these taste goals.

The development of cultured animal products has been seen as a silver bullet since its inception. In 2008, PETA announced a $1 million prize for the first team of scientists to produce large quantities of affordable, cultured chicken meat.[17] Up until the last few years, most were still skeptical that this could happen, but ever since the cultured hamburger and Perfect Day's founding, the movement has flourished with a newfound optimism, with many advocates frustrated by the seemingly slow pace of social change, believing that this new technology will cut the Gordian knot of factory farming.

A quickly increasing number of startups have been founded in this space since those landmark events. Datar and New Harvest started another one in 2014 by again connecting two aspiring entrepreneurs in their network, David Anchel and Arturo Elizondo. This team quickly applied to an accelerator, too—IndieBio in San Francisco. They initially called themselves the New Harvest Egg Project, and they utilize a similar process to Perfect Day in order to produce the roughly twelve key proteins found in egg whites. They were accepted to IndieBio, produced high-quality samples of animal-free meringues, and raised $1.75 million in seed funding before the end of the program. They now go by Clara Foods.[18]

Geltor, another company to come out of IndieBio, produces the structural protein collagen, which is found in connective tissue and most commonly used in gelatin, an animal product we rarely think of but which is a $3 billion–$5 billion industry. Alex Lorestani, cofounder of Geltor, told me he decided to focus on collagen due to not only its neglectedness but because of the industrial applications, which provide a solid, incremental model for scaling up. Collagen is used in numerous nonfood industries, such as cosmetics, pharmaceuticals, and materials, and a common complaint is that collagen from animal farming has impurities and imperfections that are costly to fix at a large scale. Cellular agriculture excels at producing a pure, homogeneous product.

Lorestani, like Renninger at Ripple Foods, designs his business strategy with an eye to the failed biofuels industry of the past decade, meaning he's wary of aiming for a big commodity market without a clear ladder of profitable business outcomes to get there. In fact, Geltor already started selling collagen in 2017. Their work, in my opinion, not only benefits their future prospects but also sets useful precedent for the food-only cellular agriculture companies that don't have the same ladder available.

Lorestani told me that his motivation was different from that of many of the other activists and entrepreneurs. He had less "moral and ethical outrage" with the issue and instead had a "technical outrage." He said that current production methods were, in scientific terms, "the dumbest way to make the things that we need and love."

There are several communities of biohackers working on smaller cellular agriculture projects, such as Real Vegan Cheese working out of Counter Culture Labs and BioCurious in California. These community research centers play an important role in providing education to young scientists and jump-starting larger biotechnology projects.[19]

The applications of cellular agriculture extend beyond the food system. Sothic Bioscience is producing LAL—limulus amebocyte lysate—an unfamiliar but important compound that the medical industry uses to test for the presence of harmful bacteria. Until a company like Sothic succeeds, LAL will be made from the blood of Atlantic horseshoe crabs with an invasive process at such a scale that it puts the species at risk for extinction as early as 2019.[20]

Other companies are applying cellular agriculture to the materials industry. The most well-known of these is Modern Meadow, which originally focused on both meat and leather. With their current focus on leather, which they see as an easier scientific and marketing project that can pave the way for food products, they use gene-insertion to produce collagen, and let that collagen combine and grow into a full hide that is tanned just like an animal's. Modern Meadow estimates commercial release in 2019, though it probably won't be selling directly to consumers.[21]

Similarly, Bolt Threads and Spiber are producing silk products with proteins made from microbes. These products are already on the market but aren't currently affordable. Bolt Threads sold a limited edition series of fifty ties at $314 each.[22]

The cellular agriculture industry today is an exciting confluence of nonprofits, engineers, biologists, and entrepreneurs working across a variety of industries with a variety of skills and perspectives.

The cultured meat arms race

There are now four main cultured meat companies vying to reach hungry consumers. The first is MosaMeat, the company founded by Mark Post and Peter Verstrate, whom we met earlier in the chapter, to commercialize their research results.

Second is Memphis Meats, the most well-known company today thanks in large part to their viral videos of prototype meatballs, chicken tenders, and duck à l'orange strips released in 2016 and 2017. I first met Memphis's cofounder and CEO, Uma Valeti, in 2015 at the Effective Altruism Global conference at Google's headquarters in San Francisco. I was presenting on effective altruism for animals and Valeti was giving a talk on cultured meat, which many effective altruists regard as a highly cost-effective way to do good in the world.

Valeti is a prominent leader in cellular agriculture with an exceptionally level-headed approach. He studied at one of India's top medical schools, then applied for a visa to finish his education at the famous Mayo Clinic in the US. He was rejected several times, so he went to study in Jamaica with plans to practice in the UK. There, he met his future wife and finally had his visa accepted on his seventh try. He moved to the US and pursued a career as a cardiologist, while his wife worked as a pediatric eye surgeon.

Valeti has wanted to help animals since he was a child. At the age of twelve, he remembers attending a friend's birthday party in India where celebration was happening in the front yard with dishes like chicken tandoori while animals were slaughtered in the backyard to feed the happy guests. Valeti refers to this as his "birthday, death day" experience when speaking to supporters and journalists. When he started learning about stem cell treatments as a cardiologist, his mind kept drifting to animal agriculture and the possibility of growing meat outside of an animal's body. He contacted Matheny and eventually joined the New Harvest board of directors in 2013. He helped search for people to take up the project of cultured meat, until finally his wife and kids told him, "You've been talking about this for a while.

You're asking other people to do it. Why are you not doing it?" So he moved to Silicon Valley and started a cultured meat company.[23]

One of Valeti's cofounders is Nicholas Genovese, who had previously worked in cultured meat research thanks to a grant from PETA. Genovese also had experience raising poultry on his family's farm. The other cofounder is Will Clem, a tissue engineer who left that career in 2012 to found and manage a Memphis-area restaurant chain called Baby Jack's BBQ. His family founded the Whitts Barbecue chain of over forty restaurants in Alabama and Tennessee, so Clem is well-positioned to help Memphis Meats commercialize their products in the Memphis barbecue scene.[24] As discussed in the previous chapter, it's critical for new animal-free foods to target the masses of American consumers, not just the mostly liberal, high-income, ethics- and health-focused consumers who are already eating animal-free food.

In June 2017, the plant-based food leader Hampton Creek announced it was joining the race for cultured meat. The company had actually been working on cellular agriculture for over a year, a fact well known to researchers and advocates in the field but kept secret until the company had more confidence in their process and a timeline for commercial launch. In fact, the company told journalists the first products would be out in late 2018, a remarkably short schedule. Memphis Meats had previously discussed 2021 for the release of its first products, and only in the same month as Hampton Creek's announcement did it push its timeline forward to 2019.[25]

Each company is vying for first-mover advantage, securing distribution channels, investor support, and public attention. In this new market, one key factor will be selling technology to and working with the big meat and food companies. As I've discussed, traditional food companies might be averse to innovate themselves due to the novelty of cultured meat and the speculative timelines of cultured meat technology. However, it seems far more likely that they will dive in once products begin hitting the marketplace. Reaching the marketplace and meeting these deadlines are actually easier than most people think. It could be one product, sold at just one restaurant, for just a few days with a high price tag, being sold at a loss for the company.

When it comes to actually scaling up, in the same way big food companies have an advantage over startups in post-innovation production and distribution, Hampton Creek has this advantage over the newer, less-established companies like Memphis Meats. Perhaps most important, Hampton Creek's plant-based R&D has given them tens of millions of dollars' worth of infrastructure in testing and cataloging thousands of plant ingredients. I discussed this work with Hampton Creek's director of cellular agriculture, Eitan Fischer, who agreed, saying that if he wanted to build a brand-new company to work on cultured meat, his first project would be to build exactly this infrastructure. The main purpose of this is to develop an inexpensive, highly effective growth media for the cell cultures. Fischer explained that all the cost models of companies in the field point to media as the limiting factor for large-scale cultured meat production.[26]

Fischer is a long-time effective altruist whom I met while he was completing his law degree at Stanford, and he echoes the dedication and ambition of Hampton Creek founders Josh Balk and Josh Tetrick. He's a quiet, intellectual young person who originally planned to pursue either corporate law, intending to donate most of his salary to animal charities, or academia, so he could share important ideas like animal ethics with law students. However, when he realized the promise of cellular agriculture, Fischer pivoted 180 degrees to a career in food technology. He saw that Hampton Creek could help push cultured meat to market years ahead of schedule, quite possibly sparing billions of animals. (Before all of this, Fischer cofounded Animal Charity Evaluators, the effective altruism research organization where I used to work.) He has never sought the spotlight, but his dedication and talent has made huge strides for the movement.

Hampton Creek is focusing on cultured poultry, largely because poultry farming has one of the greatest ethical costs. In the long term, Fischer says they hope to build a "multi-species, multiproduct platform spanning the entire range of meat and seafood."[27] If I had to place a bet on which company will be first to bring cultured meat to market, and to be its forerunner as the industry expands, Hampton Creek would be my first choice. Another possibility is that Tyson Foods will swoop into the market and take the lead, though that would likely be done through a partnership with a company like Hampton Creek. It could also be another food giant like Perdue. In fact, Josh Balk used to investigate and expose animal cruelty in Perdue's chicken meat supply chain, but by 2017, he was in talks with Perdue about how the company can invest in cultured meat.[28]

The fourth cultured meat company is the brainchild of the animal rights revolution in Israel. The amazing progress seen in Israel in the last few years is commonly discussed in the global movement, though it's not entirely clear what they did to be so successful. Based on my conversations with Israeli advocates, it seems the revolution kicked off in 2011 when two animal advocates, Daniel Erlich and Hovav Amir, added Hebrew subtitles to an online video of a presentation by American animal advocate Gary Yourofsky. The clickbait-titled video, "The Best Speech You'll Ever Hear," was of a college presentation at Georgia Tech in 2010. Yourofsky is known in the animal rights movement for his bold language with explicit comparisons of slavery, Jewish persecution, and animal farming—even calling slaughterhouses "concentration camps." The two young advocates who subtitled Yourofsky's video worked hard to make it go viral in Israel. They handed out literature on the streets, shared the video through existing activist networks, and even persuaded a tofu company to print a link to Yourofsky's site on their packaging. The movement rocketed into mainstream discourse thanks to that viral video. Popular food critic Ori Shavit went vegan and became a figurehead for the movement. Vegan activist Tal Gilboa won the Israeli version of the *Big Brother* reality TV competition.[29]

Beyond that, why has the movement taken flight so rapidly in Israel? Israel seems like a country primed for animal rights messaging. Israeli advocates think the history of persecution helps people immediately sympathize with animal suffering. Their culture praises bluntness and strong language more than Western Europe and the US. The most common religious practices involve numerous dietary restrictions, unlike the "I'll eat anything" culture of many other countries. Once Yourofsky's speech kindled the activism networks organized in the prior decade, the resulting forest fire spread quickly through Israel.[30]

In 2013, when Israeli advocates heard of the cultured hamburger, some with biology and engineering backgrounds knew they had to get into cellular agriculture, and other young advocates entered those fields specifically for this reason. Some founded a nonprofit similar to New Harvest called the Modern Agriculture Foundation (MAF) in 2014. They worked with experienced Israeli tissue engineers to evaluate the feasibility of cultured chicken, chosen because of the particularly negative impact of chicken farming on animal welfare.

With the results of their feasibility study, MAF helped found a new company called SuperMeat. As of 2017, SuperMeat had made the most optimistic predictions of any start-up in the field, suggesting that it will soon have viable plant-based serum alternatives and can produce a whole chicken breast—not just individual muscle fibers.[31]

While I'm a bit concerned that the name SuperMeat sounds more like a video game than a food innovator, I'm excited for the company's prospects because it has the weight of the Israeli animal rights movement behind it. SuperMeat ran a very successful crowdfunding campaign to signal that many Israelis are eager to buy its products, which led to significant investor funding.[32]

There are two other Israeli meat companies getting started as I write this book: Future Meat Technologies and Meat The Future. Both are largely flying under the radar, but from off-the-record conversations I've had with Israeli entrepreneurs and scientists, it seems all three are serious competitors in the Israeli race to commercialize cultured meat.

Another emerging cultured meat company, Finless Foods, focuses on seafood, a particularly neglected and harmful industry on animal welfare and sustainability grounds. Finless is still small at the time of this writing, but as the leading company in the field of seafood, they could ride the waves of success and hype that companies like Hampton Creek are generating for poultry, beef, and pork. Fish meat also has some technical and business advantages: It can be grown at room temperature, which reduces energy usage since land animal cells need to be kept at around body temperature. It doesn't need carbon dioxide regulation. It has a simple, homogeneous structure. There's also a sizable market for high-end seafood with a consumer base that's eager to experiment, especially given the adventurous palates of sushi aficionados.[33] Of course, the companies focused on land animals, like Hampton Creek, could expand into the fish market with the infrastructure they've already built.

With the rapid growth of cellular agriculture companies, the work of the Good Food Institute, New Harvest, the Modern Agriculture Foundation, and Food Frontier (a new group currently being founded in Australia with a similar approach) becomes even more important. For-profit corporations will inevitably be tempted to undercut each other for individual gains, and even to undercut their own long-term prospects in exchange for the short-term gains of investors and other stakeholders. For example, I've already heard of one animal-free food company where executives and investors driven to reap short-term profits seem to have crippled the company's long-term research agenda. Nonprofits are essential to tame and coordinate these powerful companies to ensure that changes are sweeping and sustainable.

What's in a name?

The biggest strategic debate so far in the field of cellular agriculture has been over what to call the product. Scientists originally used the term "in vitro meat" as a scientific specification that it's meat grown outside of an animal's body. Some journalists used this, but many preferred the provocative term "lab-grown." Unfortunately, "lab-grown" is quite misleading and widely disliked in the field. All new foods are originally made in labs when they're produced by food scientists and culinary professionals in testing facilities. It also isn't an accurate term to describe how the commercial product will be produced—the small scale of lab production would make any product prohibitively expensive. In reality, the production facilities will be more like beer breweries with large sealed tubs that mix cells and nutrients with the appropriate conditions for tissue growth.

Of course, "lab-grown" is also unpleasant and unappetizing because people typically want foods that sound as natural and garden-fresh as possible. As I discussed in the context of plant-based foods, terms do more than literally describe a product. Calling it "lab-grown" would stack the deck against the new product despite its huge upsides. Unfortunately, early proponents of the product did little to push back against the term—which, from the journalists I've asked, was used because it was the most attention-grabbing—until the Good Food Institute was founded and advocated aggressively against it. Therefore, some journalists have grown accustomed to using the term without pushback. It's possible that the use of "lab-grown" contributed to the excitement and investment that we see today, but I think that would have happened anyway because it's still a provocative, exciting technology regardless of terminology.

The third term, and the most common one since the scientific community settled on it in 2011, is "cultured meat."[34] This seems like an accurate term, describing that the meat cells are grown in culture, but it's still not perfect. For some, it still has the connotations of a scientific laboratory, and it has an alternative meaning in the culinary world, referring to food made with fermentation such as yogurt or beer. While the production facility might look similar to that of these products, there's no fermentation involved. This is especially confusing when it comes to dairy products. Does cultured milk mean milk made using microorganisms, or does it just mean yogurt and cheese, which are made by adding cultures to milk?

The fourth and final option, in use since at least 2008 and heavily favored by the Good Food Institute, is "clean meat."[35] This term doesn't tell you anything about the process itself, but like the term "clean energy," communicates that the product is more ethical and sustainable than the conventional option. It also communicates the food-safety benefits, as tissue-engineered meat doesn't require antibiotics, hormones, or the risk of foodborne illness common with conventional meat. "Clean meat" seems like a promising option when you consider that conscious consumers mostly care about the ethical and health impacts of their food, using the processes of food production as proxies.

But "clean meat" isn't perfect. To some it makes the new industry seem like it's trying too hard to put a positive spin on its products instead of being as transparent and open as possible by presenting only the facts. It could also sound overly trendy, associated with the controversial health fad of clean eating. New Harvest differs from the Good Food Institute here and favors the term "cultured meat."[36] Some, like Mark Post, also worry that the term "clean meat" could alienate companies that currently sell conventional—implicitly unclean—meat, though this concern is mitigated by the investment of the animal agriculture giant Cargill in Memphis Meats in 2017 and by other unpublished negotiations and deals with animal agriculture companies.[37] Considering the importance of getting big companies to produce and sell these products, whether these companies see cultured meat as an upgrade or a competitor seems like a legitimate concern. Two other arguments against "clean" are (1) that it could put too much focus on the sustainability and environmental benefits, which, as I'll discuss in chapter 9, might not be ideal in the long run; and (2) that it is artificial-sounding and off-putting, like the words "antiseptic," "sanitized," and "disinfectant." These terms are desirable in other contexts, like housecleaning or surgery, but they sound unappealing when it comes to food.

Other researchers and I tested another potential downside in 2016 with a randomized controlled trial. We showed research subjects a news story introducing either the term "cultured beef" or "clean beef," then presented a series of purchasing decisions. Each time, we described a cultured/clean product and a conventional product, both either chicken or beef, with separate randomization of the price and food type, such as one being hot dogs and the other being burger patties. Participants then decided which of these products seemed more appealing, which gave us a sense of which term had more immediate consumer appeal. The purchasing decisions in the "clean" group were 52.4 percent in favor of the clean product, while 41.4 percent in the "cultured" group were in favor of the cultured product. This suggests that "clean" was more appealing than "cultured" in this (admittedly limited) context.

These figures weren't really a surprise, but consumers' immediate reactions are just one factor in the overall impact of a product's name. We also wanted to test whether "clean" might be more vulnerable to criticism. If consumers see advocates as putting a positive spin on a product with the term, and then see that product criticized, will they feel misled and much less inclined to purchase the product?

To test this, we showed participants fake news stories. The "clean" one was titled "'Clean Meat' or 'Unclean Meat'? Critics Bash Activists for 'Misleading' Term." The article in the "cultured" group was written a little differently, since an "uncultured" criticism wouldn't make as much sense, but it was also written as a hit piece. After this, we gave participants another set of eight purchasing decisions. This time, the percentages of favorable responses were 40.0 percent for "clean" and 33.2 percent for "cultured." This was a larger drop for the "clean" group, but not so much larger that "cultured" became the more favorable term. Overall, most researchers saw this as evidence in favor of "clean" because the drop was not as large as we expected.[38]

So which parties should use which terms in the farmed animal movement? "Cultured" might appeal more to scientists, at least those not in the food industry where it suggests fermentation, while "clean" might appeal more to the general public. And, in the long run, if we want to just use the unaltered term "meat" to clarify that these products are identical to animal-based meat, then using multiple terms now could make that transition easier as people are less locked into a specific name. Of course, we shouldn't make this transition too soon, as that could mislead or at least confuse consumers.

Currently, I think it makes sense for New Harvest to use "cultured," GFI to use "clean," and for other parties to follow suit based on whether they work more with scientists or the general public. I would also like to see a transition to using "clean" to refer to both plant-based and cultured foods, in the same way "clean energy" refers to solar, wind, and other energy sources that are cleaner than the fossil fuel baseline. I typically use it this way, and refer to tissue-engineered meat as "cultured" or "cell-cultured." I suspect that people feel less resistant to using a positive term for a diverse group of products than for a single type. Overall, I still mostly use "animal-free" to refer to all products made without animal farming, as I have done throughout this book, as it is both factual and avoids any strong connotations, but I still think "clean" deserves a place in the movement's rhetorical toolbelt.

New Harvest and the Good Food Institute also differ somewhat in their approaches to information-sharing in the cellular agriculture space. To date, New Harvest has focused more on academic and nonprofit research, the results of which are frequently published publicly and not protected by intellectual property rights. The Good Food Institute has instead focused on corporate research, which often sacrifices openness for the increased profitability of private research. After all, how many investors want to fund research that could easily be used by a competitor? I'm honestly not sure which of these approaches we need to prioritize more right now as they both have solid supporting arguments. I do think that we can expect the open-source research to be more neglected, since more people are motivated by profits and social impact than social impact alone, which makes me weakly favor the open-source approach.

In the same vein, New Harvest has longer timelines for when they expect cellular agriculture products to hit the marketplace, and its staff have been critical of the short timelines expressed by the Good Food Institute and startups.[39] Short timelines make sense to drum up investment and news coverage, but too much focus on them can lead to a frustrated public and investor community if progress is delayed. This is similar to what has happened in the field of artificial intelligence (AI) over the past few decades, where slow progress has led to periodic "AI winters" of public pessimism and decreased funding. I think we need to focus on longer timelines. This is also because I expect the landscape to drift toward an excess of optimism because of the incentives of for-profit companies.

The future of cultured meat technology

Scientists need four subtechnologies to make cultured meat cost-competitive with conventional meat:

- Cell culture media, which must contain the nutrients and growth factors necessary for abundant cell proliferation. A cost-efficient media system will likely recycle its contents batch by batch, so materials not used by the cells aren't wasted.
- Bioreactors, which need to hold the cells and culture media during the growth process. Optimal conditions such as temperature and media contents need to be maintained, likely with the help of sensors.

- Cell lines—a reliable cell bank of the required type—which need to be available at low costs for the various types of meat being produced, and ideally they'll have the ability to grow into the different components of meat: muscle, fat, connective tissue.
- Scaffolding, which needs to support the cells during their growth. This is analogous to the extracellular matrix in an animal's body. For minced meat, this could be molecules distributed in the culture media, but for whole tissues like steak it might need to look more like a criss-crossed, three-dimensional network of fibers.[40]

The particular hurdles and priorities of the industry might change as the technology develops, but each of these components will be required in some form unless the tissue-engineering process changes dramatically.

I discussed in chapter 3 how companies like Hampton Creek are keeping prices low while they reach economies of scale, but a higher initial pricing model is probably better for products with more complex production methods, such as cultured meat, where lowering production costs is a more long-term and uncertain goal. As such, we'll probably see it first sold at high-end restaurants and specialty grocers. The first cultured meat products will also probably be ground beef, chicken nuggets, and similar homogeneous meats that are easier to produce. The technology will face significant regulatory hurdles, but its framing as more ethical and sustainable, as well as the support from big food companies, will likely carry it to market with relative ease once the technology advances.

In fact, there's a reasonable argument to be made that technology alone will end animal farming. Even if scientific progress for cultured meat slows down due to limited funding or unexpected technical roadblocks, it will continue if for no other reason than the inability of livestock to feed the increasing human population. As we've mentioned a few times so far, animals are simply inefficient producers of flesh, so an artificial process should be able to win out in the long run. Moreover, technological progress doesn't face the unique challenge of social progress: public opinion can move forward and backward, but barring exceptional circumstances like civilizational collapse, technological capabilities only move forward. This means that by working on technology, we usually are just affecting the speed of progress, but by working on social change, we can affect the direction of progress. Affecting the direction ensures that, in the long run, the welfare of humans and animals is as good as it can be. Consider that the long run could be millions or more years of a huge civilization stretching across the stars, while the speed of progress mostly just affects what happens in the next century.

This suggests that we might want to spend our limited resources on the social aspects of animal-free food adoption—ensuring that these technologies are not just developed, which seems fairly inevitable, but widely adopted. While society-wide adoption seems likely, especially to people like me who see the huge ethical benefits, we can never be that certain about social change. What if the big meat companies currently investing in the space decide it's easier to squash their competition than to reinvent their supply chains? What if people recognize the efficiency of cultured meat, but not to the extent that they believe it justifies transforming the food system? Social change is even more important if we zoom out and consider issues other than animal farming: if society does end animal farming with technology and for the sake of efficiency alone but fails to develop compassion for farmed animals and other sentient beings, then our descendants might inflict similar abuses in the future, comparable to factory farming. Such tragedies could be even greater in scale, as future humans will likely have greater technological capabilities.

Social change can also inspire more people to work on beneficial technologies—like the vegetarians and animal advocates who went into cellular agriculture research and development—and it could speed up the adoption of those technologies. Hastening the end of animal farming by just a few years, months, or even days could spare billions of animals from factory farms and slaughterhouses.

For these reasons and others, I'll now turn to social change as the second key to the end of animal farming.[41]

6. THE PSYCHOLOGY OF ANIMAL-FREE FOOD

LET'S BEGIN the discussion of social change with the perennial vegetarian question: With so many good reasons to eat animal-free foods, why is the (animal-based) meat habit so hard to break?

The Four N's

What do people typically say when we ask them why they eat meat? Psychologists categorize people's reasons into the four N's: meat is normal, necessary, nice, and natural. It's normal because most people are meat eaters. It's necessary for good health. It's nice because, well, it just tastes good. And it's natural because humans have been eating meat for tens of thousands of years. When consumers are asked to list three reasons why it's okay to eat meat, these categories cover around 80–90 percent of responses.[1] So how can advocates address them?

NORMAL A wide range of research has shown that social pressure strongly motivates behavior across a wide variety of contexts from environmentalism to teenage drug use.[2] Famous experiments have documented the power of social pressure, such as the Music Lab experiment of Matthew Salganik, Peter Dodds, and Duncan Watts. They split participants into two groups, both of which evaluated forty-eight different songs by indicating "I hate it" or "I love it." Both had the option of downloading the song in addition to their binary assessment. In one group, participants were shown the number of times the song had been downloaded by previous participants, while the other group lacked that information. The results suggested that the download count had a large effect on assessments of the music, while the intrinsic features of the music seemed to have little effect. This helps explain why it's so hard to predict success in the music industry just by listening to a song. It might be the case that hits reach the top of the charts by snowballing with their initial popularity, more so than by having the perfect tune and lyrics. In the words of the authors, "The best songs rarely did poorly, and the worst rarely did well, but any other result was possible."[3] This work follows psychology experiments by Solomon Asch and Stanley Milgram showing similarly powerful effects of pressure from peers and authorities.[4] The tremendous power of social pressure, from both authorities and peers, has been implicated in many historical tragedies, such as the Salem witch trials and the Holocaust.[5]

But social pressure is just a tool, not necessarily a weapon. For example, if you can get a few kids to engage in better behavior in a class or peer group, such as doing their homework or avoiding smoking, the others will be more inclined to follow. Such influence can also be more artificial: if the students have a young babysitter or teacher who acts cool, such as by liking the same music as the teenagers, but then also espouses good values, this can serve as social pressure to improve behavior.

Farmed animal advocates use similar tactics, such as by highlighting the growing number of vegetarians and vegans and by profiling celebrities who care about animals. Mercy For Animals ran Facebook ads that specifically targeted fans of Ariana Grande, a popular singer, to tell them that Grande is vegetarian.[6] Israeli activists organized marches advertised as having ten thousand attendees, even before they had anyone signed up. These marches became self-fulfilling prophecies where the apparent popularity of the event led to actual popularity.[7] Strategies like these appear to greatly increase advocacy effectiveness and should be utilized more often.

In the long run, government and corporate policy can help tackle the "normal" objection by changing the default options for consumers. When a company orders lunch for its employees or an airline provides meals to its guests, it can offer animal-based food as a special request with animal-free food as the default. This indicates that animal-free is normal in addition to the direct impact of increasing animal-free food consumption because people are always hesitant to deviate from the status quo.

NECESSARY "Necessary" is the most common factor expressed in self-report. It makes up around 35–40 percent of consumer responses to the question of why it's okay to eat meat.[8] Fortunately, this is the most easily refuted justification. Advocates can simply share the facts of the situation: Many vegetarians and vegans are healthy, and indeed tend to be so at even higher rates than the general population. There are vegan Olympic athletes, bodybuilders, mothers, children, and grandparents who have been vegetarian and vegan for decades. While the evidence here is mostly observational studies, experiments lacking complete methodological integrity, and theoretical speculation based on human biology—and that's unfortunately all we can work with in modern nutrition science—major nutrition and health organizations agree that reducing one's consumption of animal products is not only a healthy option, but a healthier one. Kaiser Permanente, a major US insurance provider, is even encouraging their customers to adopt plant-based, whole-food diets to save money.[9] There's simply no compelling reason to think animal products are necessary for good health.

NICE While a lot of people enjoy the animal-based foods they're used to eating, we can tackle this justification by developing and showcasing tasty animal-free foods. Many people are pleasantly surprised when they learn that a delicious dessert they had was actually vegan, or when they see foods like cheesesteak at a vegan restaurant. And a simple Google search for vegan versions of our favorite recipes will often turn up a range of enticing options. I've known many people who gave vegan food to close friends without telling them. Their friends usually never guess it's animal-free, and the "nice" justification very quickly dissipates.

Of course, this doesn't change the fact that many people genuinely prefer animal-based foods to most animal-free foods available to them. In those cases, we need the increased availability and development of animal-free foods that match animal-based foods in culinary profile.

NATURAL Finally, we have to tackle the "natural" objection, which is probably the second-most-powerful psychological roadblock of the four N's after "normal." The belief that eating meat is natural is especially tenacious because it's based on fact: humanity has farmed or hunted animals for most of its evolutionary history. We'd need a drastic reduction in the scale of animal farming to match our ancestors' diets, however, because modern humans probably eat far more animal products than our ancestors did.[10] One could argue that animal-free meat, dairy, and eggs can substitute for the small amount in our ancestral diet, but these new foods are produced with unnatural processes, and one study actually suggested that consumers care more about processes than ingredients when assessing naturalness.[11]

Because new animal-free foods like cultured meat aren't natural in the typical sense—meaning, roughly, that they wouldn't exist without modern human technology—should we abandon them?

To definitively conclude that something is good because it's natural is known by philosophers as an "appeal to nature," a well-known logical fallacy. It's simply false that all natural things are good and all unnatural things are bad. Murder, rape, abuse, and many other cruelties have existed through most of our evolutionary history. Indeed, many of the boons of modern civilization such as medicine and natural disaster relief are defenses against the unfavorable natural state of humanity. There may be a lot to appreciate in the natural world, but we have vaccines, refrigerators, toilets, and air-conditioning because they're better solutions to our needs than what nature gave us.

We should also keep in mind that virtually no modern food is natural by this definition. In terms of ingredients and raw materials, modern farmed animals are the result of aggressive artificial breeding that makes them produce many times the amount of meat, dairy, or eggs they would naturally. Chickens raised for their meat today are over four times larger than they were in the 1960s, and in fact virtually all living organisms used for food, including plants, have been selectively bred to improve their quality, edibility, or productivity. Even on the smallest and most pristine farms, modern crops like bananas and corn are virtually unrecognizable when placed next to their natural counterparts. In fact, a 2017 study suggested that pointing out how consumers already eat many "unnatural" foods assuages their concern about the unnaturalness of cultured meat.[12]

The crops fed to humans and farmed animals are grown with numerous pesticides and herbicides. These chemicals are more likely to be natural than synthetic on organic farms, but as an article in *Scientific American* referred to the distinction between them, it's a "gray matter." The article even called out the belief that "synthetic chemicals are more toxic than natural chemicals" as a common misunderstanding of organic farming.[13] Farmed animals themselves are given numerous antibiotics, hormones, and other synthetic chemicals.

Beyond ingredients and processes, even the food-production environment is quite unnatural, such as the popularity of monoculture in crop production as opposed to the diverse environments plants grew in before human manipulation. Animals are also farmed in facilities that are unrecognizable from the diverse ecosystems where their ancestors lived before industrialization, at massive scales that destroy the local soil and water. Chickens raised for meat are slaughtered at around forty-two days old, and even if they weren't slaughtered so early, they wouldn't make it to their maximum lifespan (up to fifteen years before the intense artificial breeding of the last century) without succumbing to severe health issues like heart failure and ammonia burn from the harsh air quality in chicken sheds.[14]

So, are all modern foods so unnatural that we should just throw out the term "natural" in describing them? Maybe, but the issue isn't quite so black and white. We can think of naturalness instead as a sort of precautionary principle: foods that are more natural, such as those that have existed without incident for a hundred years instead of twenty, have a track record that suggests they're less likely to be harmful in the ways that ultimately matter, such as food safety, animal welfare, and sustainability. In other words, it might be wrong to definitively conclude that what's natural is good, but there might still be a correlation between naturalness and goodness. Beyond the ethical aspects, foods that have been around for longer are more closely tied to our food traditions.

However, by these meaningful standards, animal-based foods fail miserably with higher risks of food poisoning, contamination, and allergic reaction; worse long-term health outcomes; and local pollution with an unsustainable drain on natural resources. In terms of food traditions, the reality is that we can continue having summer barbecues and turkey on Thanksgiving with the animal-free versions of these foods. Very few people feel tied to the animal origins of these foods, and if that's what really matters to them—eating the flesh of a sentient being—then I don't feel bad at all for denying them that. The vast majority of people eat animal products not because of how they're produced, but in spite of it.

As advocates, we can't rely entirely on this nuanced definition. Food companies have pushed consumers to care about the term "natural" without giving thought to what they ultimately care about, and they're able to do this so successfully because our psychology is driven by our evolutionary history, when instinctively avoiding strange and unusual foods was a useful heuristic for survival.

In other words, we're irrational consumers, and we need to account for that rather than just confront the justification that something is natural on purely logical grounds. Consider the study that found that over 80 percent of respondents wanted mandatory labeling for all foods containing DNA.[15] The survey didn't mention that DNA is the genetic code within all plants and animals. I guess this labeling would distinguish such foods from, well, salt, but it seems to indicate a general fear of acronymized and artificial-sounding ingredients. When the FDA requested public feedback on its policies regarding the term "natural," food advocates expressed serious concern about its ambiguous meaning and the tendency to mislead consumers. One petition suggested the agency ban the term entirely from food packaging.[16]

A recent example of technology that some people have felt is harmfully unnatural has been genetically modified organisms (GMOs). While we've been modifying genes through selective breeding for centuries, only in the past few decades have we been directly engineering genomes with biotechnology. GMOs have some major upsides, like saving Hawaii's papaya industry and potentially providing Vitamin A–loaded golden rice to the global poor, but they haven't been accepted as easily as other technologies.[17] Let's not dive into whether GMOs are good or bad, but on the meta-level, I think a major social cause of GMO opposition is that they were introduced to the public from behind closed doors, appearing in our food before consumers knew about them. And when the public discussion finally happened, it was controlled by anti-GMO advocates who focused on their potential risks to food safety, the environment, and local economies, instead of the potential benefits.[18]

So far, the cellular agriculture industry has taken this lesson seriously, coming out early to the public and focusing on the benefits to the economy, public health, animal welfare, and the environment. This might be due to cultured meat's origin with nonprofits and academic scientists, rather than with large corporations.

One quirky example is the Shojinmeat Project, a Japanese nonprofit organization of biohackers working to educate the public about cultured meat. They even make meat themselves, using jerry-rigged bioreactors filled with bits of meat and cell media like nutritional yeast. They also have artists raising funds by selling comic books with characters and stories related to cultured meat. They especially focus on outreach to children, who find the idea of growing meat especially fun and exciting. In the words of Shojinmeat founder Yuki Hanyu, "If cultured meat is something primary school students can do, why would you fear that?"[19] (Hanyu and his colleagues are also now working to commercialize cultured meat in Japan with a company called Integriculture.)

It's essential that we continue this trend of openness, ensuring that the public discussion of cultured products is transparent and thoughtful rather than hushed and fearful. Note that I'm not equating GMOs to animal-free food in any way here—even cultured products are not necessarily genetically modified—but both are new food technologies that have faced or could face public opposition.

Data suggests that current hesitation around eating cultured meat is largely driven by the perception that it's unnatural. In one study, participants were shown one of two messages, either describing red meat or in-vitro meat along with a warning for both groups that the food they described contributes to the risk of colon cancer. They were asked about how artificial or natural they thought the food was, and then how acceptable the risks were. The researchers found that the "red meat" group on average felt the risks were fairly acceptable, scored at 59 on a scale from 0 to 100 ranging from "not acceptable at all" to "very acceptable." However, the "in-vitro meat" group scored that food at only 25, despite the same risks being described to each group. In fact, when I read through the methodology, I saw that the "red meat" group was actually also shown information about the animal suffering and environmental harm that the "in-vitro meat" group was not shown, which makes this difference even more compelling.

Statistical analysis of the acceptability ratings combined with the answers to the question about the food being artificial or natural showed that people's perceptions of naturalness mediated their views on risk acceptability. When the "natural" question was controlled for in the results, the difference in risk acceptability was no longer statistically significant.[20] In other words, people are more concerned about the risks of cultured meat because they see it as unnatural. Because of this effect, health and food safety are even more important for cultured meat producers than they are for the conventional meat industry.

What do surveys suggest for consumer acceptance of cultured meat? Not much. Out of twenty-six surveys as of late 2017, the number of consumers who say they would eat cultured meat varies from 16 percent to 66 percent, with up to 30 percent saying maybe and the rest saying they would not eat it. It turns out that the framing of questions, especially the term used to refer to cultured meat, greatly affects the results. People are fickle, and in general they are averse to any new food technology, even if they would buy it in stores. In these surveys, just being told a few details about cultured meat in the survey has increased acceptance by 10–20 percent.[21]

All things considered, my recommendation for the farmed animal movement is this: advocates should remember that naturalness is not what people should ultimately care about; the term just approximates some of the metrics that really matter, such as food safety. Plant-based and cell-cultured foods excel by those metrics. In our activism and marketing, we should be transparent about animal-free food production so we can prove this to our audience.

In the long run, the movement can also tackle the argument that meat is natural and the other N's through government and corporate labeling campaigns. In the case of cigarettes, before labeling laws many consumers made everyday purchases without having to confront the serious health harms of smoking.[22] Similarly, consumers could benefit from negative labels on conventional meat, dairy, and eggs, and positive labels on animal-free foods featuring their health, sustainability, and animal welfare benefits. Many ideas for government policy can be found in the work on psychological "nudges," small changes in framing and wording that create big, positive changes in public behavior.

Finally, advocates should remember that the evidence is on their side. Instead of spending so much time defending against misguided concerns, keep the conversation centered on the benefits when possible. There are numerous historical examples of advocacy backfiring because it focused on dissuading misinformation about negative side effects, rather than emphasizing the huge benefits of the new system.[23]

The problems with "humane" animal farming

We need to tackle a fifth justification for animal-based food that's particularly tricky: people readily agree that factory farming is bad, and even applaud food advocates for encouraging a shift away from such farms, but many argue that some animal products come from humane farms, where the animals are treated well and live good lives. They say that if they only eat animal products from these farms, then their consumption is justified. Similar arguments are made for other "specialty farms," those supposed to be better for the environment, better for public health, or better than factory farms in another ethical aspect.

Many vegetarians and farmed animal advocates—probably most readers of this book—will agree that this justification sounds sensible. So did I until about two years ago. Now I think most advocates should oppose *all* animal farming, not just factory farms.

There are three main arguments in favor of this claim:

1. The exploitation of any sentient being—using them for one's own purposes in the way we use desks, shovels, and inanimate machines—is intrinsically wrong. It is a moral misdeed even if those sentient beings live happy lives.

2. Animals raised on specialty farms still suffer greatly, and these farms are still worse for the environment and public health compared to plant farms. By some metrics like

land use, some specialty farms might be even worse than factory farms. They could be made more ethical with additional financial costs, but those costs seem far larger than what society is willing to pay.

3. If we support specialty animal farms, even if they actually are ethical, that will lend social, political, and economic support to animal farming as a whole and therefore to unethical animal farms. For example, treating some animal farms as ethical allows people to justify their own animal product consumption by taking "psychological refuge" in the belief that they are eating ethically produced food, even if the vast majority of such consumers are not actually eating those products.

The main argument on the other side is that public figures, institutions, and the public in many countries are already happy to denounce factory farming while few have denounced animal farming as a whole. This argument suggests that an anti-factory farming message could garner more support from these sources.

Remember that just because the movement as a whole should favor one strategy doesn't mean that all advocates should use that strategy. In this case, even if a focus on animal farming seems more compelling, it probably makes sense for the most moderate voices working on campaigns like cage-free eggs to still focus on factory farming.

The wrongness of exploitation

Most of the arguments in this book are strategic, but since this one is moral, we will need to use the tools of moral reasoning like thought experiments to tease out your moral intuitions.

Picture an alternate version of our world in which Homo sapiens' cousins, the Neanderthals, surpassed us to become the most powerful species on the planet, able to dictate the fate of all sapiens and other species like sapiens do in the real world. In this world, we sapiens still have all the same capabilities, but Neanderthals are more intelligent, better with language, and better able to form large cooperative societies—in fact, they surpass us to the same extent that we surpass chickens, fish, pigs, and other farmed animals.

Let's also assume that these sophisticated Neanderthals have hunted and eaten sapiens and other intelligent primates throughout their evolutionary history. Even once they no longer need to eat us, just as we no longer need to eat chickens and fish, they continue to do so, perhaps due to taste, tradition, or the other reasons sapiens have for eating animals today. However, these Neanderthals are not as cruel as real-world sapiens. They give the farmed sapiens plenty of space, keep them in good health, and even provide them with entertainment and intellectual stimulation. The sapiens are entirely content with their living situation. When they're around fifteen years old, they're taken to a very hygienic and spacious slaughterhouse and meat processing center, and hit in the head with a bolt gun to stun them before they can smell, see, or hear what's happening, so they die as quickly and painlessly as possible. Not a single sapiens farm or slaughterhouse deviates from this standard of treatment.

So what do you think, is it okay for these Neanderthals to farm humans "humanely"? Many people will still insist that there's something terrible going on here. For many, the exploitation itself—raising and killing sapiens for Neanderthal benefit—is wrong even without the suffering. If you're in this camp, you must ask yourself: Is there any relevant difference between the hypothetical sapiens and any chickens who are exploited but otherwise raised ethically? Remember that the chickens only have lesser cognitive ability than we do to the extent that we have lesser than the hypothetical Neanderthals, so relative cognitive ability is not a plausible reason to treat the situations differently. Most people will find that this thought experiment brings to light a discomfort humans have with the connection between sentient beings and the steak or fish filet on their plate, even if those animals are treated humanely.

The humane myth

I mentioned before that in the course of my research, I wanted to get firsthand experience at a factory farm, but I also wanted firsthand experience at the most humane animal farm I could find. Fortunately, many small farms, such as those who set up stands at farmers' markets, are willing to let people visit their facilities.

So in March 2016, I drove from San Francisco up California's northern coast, through towering redwoods and past crashing waves. As I pulled up to this award-winning farm, I saw a landscape of bucolic green grass and rolling hills like the backdrops you see in the most enticing meat advertisements. To the east were a few distant peaks of the Pacific Coast Ranges, and standing in the middle of a football field-sized pasture peppered with chickens was a run-down-yet-functional mobile chicken coop. I thought to myself, why couldn't all farms be like this? I had seen what happened behind the locked doors of factory farms, but here I seemed to be witnessing a better way.

The devil was in the details. Upon closer inspection, the birds were in worse health than those of any other farm I'd been to, with numerous cases of Marek's, a highly contagious disease that had led to partial blindness in many of them; swollen abdomens, some with over a pound of fluid buildup in their less-than-five-pound body; vent gleet, a fungal infection in the opening of a bird's reproductive and digestive tract; and lice. Like the hens in factory farms, many of them suffered and some had already died from cancer, stuck eggs, reproductive tract infections, and other illnesses and ailments that come from being bred with such hyperactive reproductive systems.

The farmer told me about the challenges of pasture farming, like how many birds he lost to predators. He didn't mention anything about how it affected the birds' welfare; he just noted it as a reason their eggs were so expensive. To stem the loss, he chained a dog to the chicken coop, a Great Pyrenees with matted fur and no water nearby. The gentle giant constantly uttered soft whimpers during my visit. These eggs were priced at over $6.00 per dozen, which still excludes the taxpayer costs of farm subsidies, health-care costs, and other issues hidden behind the labels and price tags on any carton of eggs. This cost, even assuming the animals live good lives, would still exclude many consumers from purchasing animal products, especially in the large quantities they currently consume.[24]

When I visited the farm, I sincerely wanted to believe that these animals had good lives, but the evidence wasn't there to support it.

I was disappointed by this and other personal farm visits, as well as evidence from hundreds of other visits to "humane" farms by animal protection advocates and investigators. As discussed in chapter 2, some investigators have actively sought out humane-certified farms, and still inevitably find rampant suffering. One exposé, for example, found that Whole Foods was advertising the "showcase farm" of one of their turkey-meat suppliers, when that farm produced no turkeys for commercial sale and was strictly maintained for private use and tours.[25] There's a valid concern that these investigators might be biased toward seeking out the most egregious examples of cruelty, given many of these investigators view all animal use as immoral, but the abundance of evidence they have collected makes it clear that both big and small animal agriculture companies often severely distort the truth. We should also consider the bias of people who eat conventional animal products, such as most agriculture experts and journalists who communicate the results of investigations with the general public. It's worth mentioning that pasture-raised cows seem to have better welfare than pasture-based or otherwise "humanely raised" chickens or pigs.[26] Though pasture-raised cows still endure significant health issues and psychological torment from family separation, and are slaughtered at the same horrific facilities as factory-farmed cows.

Many people are also surprised at how bad these pastoral farms are in terms of human health and environmental impacts. Many of the components of animal products that health advocates find problematic, such as cholesterol, saturated fat, and carcinogens, are still present in products from small, humane, organic, or other specialty farms. There might again be a limited exception here for grass-fed beef: Studies suggest it has slightly more omega-3 fatty acids and other beneficial nutrients than conventional beef, though the potentially harmful compounds in beef are still present and those beneficial nutrients are available in animal-free foods.[27]

In September 2016, the *New York Times* ran the provocative article "Why Industrial Farms Are Good for the Environment." In it, economist Jayson Lusk argues that large farms can use sophisticated technology that benefits the farm and the environment, such as variable rate applicators, which apply only as much fertilizer as each specific area of the farm needs to avoid waste. He compares the number of animals used by modern farms with the number used by 1950s farms to produce the same amount of food. For beef, today's technology reduces headcount by 34 percent and for dairy, 76 percent. The article also suggests we use 16 percent less land for each unit of food production than we did in 1970.[28]

By some measures, grass-fed cow farming is potentially worse than grain-fed cow farming. Grass consumption leads to two to four times more production of methane, a major greenhouse gas. It also takes more land, water, and fossil fuels to produce grass-fed beef, even considering the resources used to produce the soy and grains that are used to feed conventionally raised cows.[29] Overall, some animal farms do end up with a significantly smaller environmental impact, such as farms that rotate pasture-raised cows with crops. However, specialty farming does little to mitigate health concerns aside from the overuse of antibiotics. And the animal suffering, especially on specialty poultry farms, is still staggering.

You may be thinking that even if most specialty farms still have serious issues, surely at least a few farms—perhaps even just a handful—have happy animals. Maybe you've even been on a farm that you're convinced has happy animals. This response is valid, to an extent. Every day on my commute to high school in rural Texas, I drove past pasture-raised cattle who seemed perfectly content to chew their cud. Yes, slaughter might have been a terrible experience, but it seems plausible that one day of even intense suffering might not outweigh a few years of happy cud-chewing life.

Unfortunately, there are serious limitations to scaling up that tiny number of farms while maintaining positive welfare. At an animal law conference I attended at Harvard University, Bernadette Juarez, deputy administrator of the USDA's Animal Care program, stated herself that the agency has no chance of protecting all farmed animals, even just to enforce the extremely limited regulations that exist today. She made this qualification to appease animal advocates in the audience, to explain why cruelty still happens despite USDA staff doing their best, but in doing so she inadvertently put forward one of the strongest arguments for opposing specialty animal farms: they're far too costly to feed our hungry planet.

Given the inability of farmed animals to revolt, to break out of a farm and tell the world about their abuse, we would need extensive regulations to maintain high animal welfare throughout the industry. This would include the expenses of regular independent inspections and livestreamed security footage at all facilities. Consumers or taxpayers would also need to pay for direct costs such as more space per animal, an army of veterinarians and medical supplies for sick and injured animals, and a reversion of the artificial breeding that has made these animals grow muscle and fat at ultra-fast rates and produce huge amounts of milk and eggs. That level of welfare doesn't even exist at the very best farms today, so even the steep price tag of over $6.00 per dozen eggs from the pasture farm I visited—not including externalities—is still not high enough to guarantee that the animals have good lives.[30]

So even if humane animal farming is possible in theory, it seems exceedingly difficult to achieve at a financially accessible global scale.

Psychological refuge

Every grassroots farmed animal advocate I've asked about this topic has spoken with many people who insist that the meat they buy doesn't come from factory farms. "I only eat humane meat," they say, defending themselves from the activists' critiques of factory farming. This is one of the most common justifications heard by grassroots advocates.

Is it valid? Consider that, by the best estimates we have, over 90 percent of animals in the global food system (over 99 percent in the US) are in factory farms.[31] Could it really be the case that so many consumers are discerning enough to ensure that most or all of the animals they consume are from the few specialty farms that, as we saw above, might or might not even actually be ethical? It's very unlikely. In fact, a survey my colleagues and I conducted in 2017 suggested that 75 percent of US adults say they usually consume humane animal products, which seems impossible given that the best estimates suggest less than 1 percent of US farmed animals live on nonfactory farms.[32]

This is also supported by advocates' experiences with these well-meaning people who believe the animals they eat were happy. Advocates often ask the person about their recent consumption choices: What did they have for breakfast? Scrambled eggs. Where were the eggs from? The college cafeteria. What do you know about their sourcing? They're cage-free. But that's a far cry from humane treatment.

Usually at this point, the uncomfortable consumer will bring up the few instances where they actually buy from specialty farms, such as beef patties from the farmers' market, but this is a small fraction of their total consumption, and they have done little investigation of the ethics of those farms, instead just relying on a label saying something like "cage-free" or "humanely raised." And most don't even think about other considerations, like what they eat in restaurants or whether they consume fish, virtually none of whom live on farms that even claim to be high-welfare.

Many thoughtful vegans actually got their start trying to be ethical omnivores. One college student, Jay Shooster, tried to do this by adding a farm husbandry course to his curriculum and approaching local farmers with a list of cruel practices. He asked the farmers which, if any, they employed and was disappointed by the results every time. This led Shooster to go vegan, and he now works as an animal activist.[33]

When people call upon the idea of ethical animal farming—even if that constitutes little or none of their actual consumption—we can think of it as a "psychological refuge" they're using to justify their consumption of factory farmed products. This refuge shelters them from the cognitive dissonance they would feel if they both fully considered their ethical views and the realities of their consumption choices. It's one of the biggest roadblocks to fixing our food system, perhaps even more harmful than the four N's.

When we look at successful social movements, they had simple demands. Abolish slavery. Votes for women. The antismoking movement didn't tell people to "avoid cigarettes produced by big corporations." Marriage-equality activists didn't push for "the right to marry who you want or to enter into some other legal contract with similar benefits." They kept it simple. We see straightforward messaging succeed in the corporate world, often to the point of leaving out virtually any meaningful information about a company's products—Nike's "Just Do It," for example, or Hilton's "Travel should take you places." The farmed animal movement needs a simple message too: end animal farming.

Importantly, belief in ethical animal farming also reinforces the notion that animals are mere property, inanimate beings that exist only as means to human ends. So long as other animals are considered our property, even their most desperate needs come second to our most trivial wants. It's this attitude that lets us condemn billions of animals to the torment of factory farms. Indeed, psychological research suggests that the idea of meat consumption, like research subjects being told they would soon get some beef jerky, leads people to assign fewer mental faculties to farmed animals than people prompted to think about eating a plant-based snack.[34] By reducing people's concern for these animals through specialty animal farming, this effect could perpetuate factory farming.

▪ ▪ ▪ ▪ ▪

There's a tension between the arguments I've made here against "humane" animal farming and my argument in chapter 2 that incremental reforms like cage-free eggs are important and beneficial for the food system. I think the "humane" arguments do in fact mitigate the positive impact of welfare reforms, in three ways:

1. The more one thinks the ethical harm of animal farming is intrinsic to exploitation rather than a feature of factory farming, the less of a relative impact reforms have.

2. So-called "humane" farming actually involves much more suffering than people think.

3. By praising welfare reforms, we encourage the psychological refuge of "humane" animal farming.

I believe these are valid concerns, but I don't think they reduce the benefit so much that they actually induce more complacency than momentum for the farmed animal movement. And we can't lose sight of the concrete, significant impact of something like moving a chicken out of a cage that's barely larger than her own body. With point 3 in particular, I believe we can mitigate this concern by being careful with our messaging. We should ensure that we frame reforms as just as an incremental step toward the long-term goal of an animal-free food system.

Finally, there's a substantial marketing and psychology literature on effective persuasion techniques that can be applied to animal-free food advocacy. Much of it is intuitive, such as people being more persuaded by people who are similar to them, or listening more to people who seem like authority figures. But it's quite useful to be familiar with these findings as an advocate or even just in one's social and work life. I recommend the classic text *Influence: The Psychology of Persuasion*, by Robert Cialdini. I'll discuss some of the principles of persuasion as they apply to the social change arguments made in the next two chapters.

7. EVIDENCE-BASED SOCIAL CHANGE

ADVOCATES CAN'T BE SATISFIED with merely changing individuals one by one. To make significant progress, the movement will need to consider not just the psychology of the individuals involved, but also the structures and relationships that connect people and determine their collective behavior. Considering broader systems enables us to create change at a much larger scale.

For example, if revolutionaries want to overthrow a dictator, it's not enough for the majority of individuals to want the dictator gone. They need to know that other people want this result as well. With that knowledge, the average disgruntled citizen is able to speak out because they know they can count on support from their compatriots. To achieve this state of collective understanding, referred to as "common knowledge," revolutionaries can do something like stage a large protest. If they can get a large enough crowd at the beginning—perhaps a group of gutsy, idealistic young people willing to take action without the guarantee of popular support—then the assembly offers safety in numbers that could draw in further participation. In a similar way, farmed animal advocates seek to not just convince individuals that they shouldn't eat animals, but to show them that other people are also convinced of these arguments, which could help us achieve progress faster than one-by-one change.

Institutions over individuals

I'm often asked what the most important single lesson is from advocacy research. Which one conclusion, if the movement took it to heart, would lead to the biggest increase in our impact? While there's a lot of uncertainty here, my answer is that the movement should shift away from the current heavy emphasis on individual change and focus more on institutional change.[1]

The most common concrete example is a change in the call to action we give our audience after telling them about the issues of animal farming. The most common asks are "go vegan" and "leave animals off your plate." These are both individual change messages, and they have been the status quo in the farmed animal movement since its inception. Even *Animal Liberation*, credited as the founding text of the modern animal rights movement, has a heavy focus on individual diet change as the solution to reduce animal suffering in the food and biomedical industries. Today, we should shift toward institutional messages that focus on changing companies, governments, and social attitudes with statements like "End animal farming," or "America needs to eat less meat." These call for broader, collective solutions.

Prioritizing institutions over individuals also means pursuing different interventions. Advocates focused on changing individuals may hand out leaflets and donate to organizations that run vegetarian ads, such as one on Facebook featuring a picture of a suffering animal with the text "Click this video to find out the shocking truth behind the meat industry." Institution-focused interventions are things like petitions calling for McDonald's to end its use of battery cages, political initiatives to have more animal-free food in colleges and hospitals, and op-eds that address the urgent need for the world to switch to animal-free foods. Petitions aim to force companies to change their policies, political initiatives aim to compel governments, and op-eds aim to create common knowledge among readers and shift social norms.

The current heavy focus on consumer change seems to have numerous negative consequences for the movement. As a brief illustrative example, a few weeks ago I was at a protest for a restaurant chain to adopt welfare reforms for the chickens in its supply chain. A pedestrian walked up and thanked me profusely for our activism. When I suggested he join us—as bystanders sometimes do—he said, "Oh no, I'm not vegetarian." I insisted that he could join us without being vegetarian, but he was unconvinced. He saw vegetarianism as a prerequisite to helping farmed animals, likely due to the strong conflation of those two ideas in public discourse. If the movement had instead framed helping farmed animals as a collective issue with collective solutions, we might not have missed out on a potential activist for our demonstration.

Survey, voting, and lifestyle data

In October 2017, colleagues and I asked a census-balanced sample of US adults for their attitudes toward seemingly radical changes in food policy. Respondents were remarkably supportive: 49 percent supported "a ban on the factory farming of animals," 47 percent supported "a ban on slaughterhouses," and 33 percent supported "a ban on animal farming." When we published these results, there was widespread disbelief. How could we get these huge figures if only around 2 percent of US adults rigorously follow a vegetarian or vegan diet and less than 10 percent self-identify as vegetarian?[2]

The answer is that people are far more willing to support institutional change than they are to change their individual consumption. We've already seen this in the farmed animal movement when it comes to welfare reforms. US adults consistently show over 70 percent poll support for various changes in farmed animal welfare, such as cage-free, slower-growth chicken genetics, higher-welfare slaughter methods, and an end to extreme crowding.[3] There have also been consistent majority votes in favor of farmed animal welfare ballot initiatives.[4] This widespread support contrasts with the tiny number of consumers who actually opt for these higher-welfare products in their individual consumption: organic meat made up just 1.5 percent of conventional fresh red meat sales in the US and grass-fed 0.9 percent in 2016.[5]

Our 2017 poll also found that a whopping 97 percent of respondents agree with the statement "Whether to eat animals or be vegetarian is a personal choice, and nobody has the right to tell me which one they think I should do." I cannot stress enough how resistant people are to individual consumer change, especially when it's as closely tied to personal identity as vegetarianism and veganism are in the US public consciousness.[6]

Historical precedent

The farmed animal movement has a virtually unprecedented focus on individual and consumer change. It's difficult to find even portions of past movements with a comparable emphasis. One of the few was the free produce movement, a contingent of antislavery activists who focused on individual abstinence from slave-made goods. Similar to veganism, this was seen as reducing the economic power of slavery, signaling opposition to slavery, and helping consumers disassociate from an immoral institution. This approach was most popular in the early 1800s in the United States. The prominent abolitionist William Lloyd Garrison boasted at the 1840 World Anti-Slavery Convention that his suit was made by nonslave labor. However, by 1850, the movement lost much of its momentum once activists—including Garrison himself—decided there were more effective ways to fight slavery, such as public protest and legislative activism.[7]

Some environmental advocates feel the same way about "green consumerism," the heavy consumer focus often seen in the environmental movement.[8] One popular argument against green consumerism has been that it makes potential activists complacent and therefore less likely to work on bigger changes like environmental policy efforts. There is some empirical evidence for this effect—known as "moral licensing" or "moral fatigue"—but the research is limited.[9] It does seem like this movement has been shifting recently toward institutional, campaign-oriented messaging (e.g., "move beyond coal") and away from both individual messaging (e.g., "please recycle") and vague, broad institutional messaging (e.g., "save the earth").

One potential counterexample to the historical precedent for institutional messaging is the success of the antismoking movement, which mostly used individual messaging in public outreach. The strength of this counterexample depends on the relevance of different movements. Is the farmed animal movement more like the movement against smoking or more like environmentalism? One key difference between antismoking and animal-free food advocacy is in their respective motivations: the main impetus to shift toward an animal-free food system is usually either animal welfare or environmental concerns, while the main impetus behind antismoking has almost always been consumer health.[10]

The collapse of compassion

The enormity of animal farming and the suffering it causes sometimes leads both the general public and advocates to suffer from the "collapse of compassion." In social psychology, this refers to the relatively low levels of compassion people tend to feel for big problems that affect many individuals. The leading explanation for this collapse is that "people expect the needs of large groups to be potentially overwhelming, and, as a result, they engage in emotion regulation to prevent themselves from experiencing overwhelming levels of emotion." A key way to mitigate the collapse of compassion seems to be to show a realistic, collective solution to the large problem.[11]

Institutional messaging does this by emphasizing the possibility of significant progress beyond what an individual can achieve by changing their own diet. Even though a vegetarian diet can spare 371 to 582 animals each year, that can feel like a mere drop in the bucket when people think about the billions of animals who are still suffering.[12] On the other hand, collective action can achieve technological breakthroughs, corporate and legal policy change, and society-wide shifts in consumer behavior. This collective approach also makes individual action easier because each person knows that their action is part of a wide movement with support and cooperation to achieve their goals.

In grassroots activism, the benefits of collective action are undeniable. One of the founders of the field of sociology, Émile Durkheim, wrote in 1912 about "collective effervescence," the powerful psychological effect of people coming together, performing the same actions, and thinking the same thoughts. As people dance together at a deafening concert, recite a religious text in church, cheer for a favorite sports team, or protest against a common enemy, they lose their sense of individuality and gain a sense of "sacredness." This can satisfy our psychological need for belonging and spirituality, potentially allowing us to break from the status quo of our individual behavior and beliefs.[13]

To critics, a collective shift to an animal-free food system might seem so implausible that a focus on individual diet offers a more inspiring solution despite its small scale. They could also see institutional messaging in general—calling for everyone in society to change their behavior—as too totalitarian or aggressive for people to accept.

I think this is a valid concern, but not so damning that we need to maintain the overwhelming focus on individual diet change that we currently see in the movement. The end of animal farming is possible, especially given recent advances in food technology such as the world's first cultured beef and poultry, and the need to feed a growing global population. Mainstream media outlets and even animal agriculture industry publications have repeatedly acknowledged the possibility.[14]

Moral outrage

Moral outrage is "a special type of anger, one that ignites when people recognize that a person or institution has violated a moral principle (for example, do not hurt others, do not fail to help people in need, do not lie) and must be prevented from continuing to do so."[15] It's the nitromethane of a social movement, exploding in the engine to power it over the finish line. Moral outrage inspires average citizens to step onto the streets, call on their loved ones to join the movement, and step outside of their comfort zones. Anger in general, including moral outrage, is cited as a key motivator by activists in many movements.

Moral outrage has also been described as "a response to the behavior of others, never one's own,"[16] so institutional messaging is more likely to spark the emotion because it puts the blame for the ethical issues of animal farming on the industry, government, or society at large, rather than on the individual. Because of this, institutional messaging could face less of the defensiveness advocates frequently encounter.

A key aspect of moral outrage is that it seems to make people more willing to break from "system justification," the tendency to justify the status quo, often for no other reason than because it is the status quo. This argument makes sense in theory and has some empirical evidence.[17] Given how common system justification is when people hear "go vegetarian" messages, feeling an immediate urge to justify their own diet, this could be a very important effect of institutional messages. It could reduce the number of irrational arguments that advocates hear, like the infamous "Lions eat meat. Why can't I?"

In addition to evoking more moral outrage, institutional messaging inspires the audience to take the issue more seriously by presenting it as sufficiently important for major institutions and society as a whole to take action. Consider that when people in the animal agriculture industry want to dismiss or minimize the messages of farmed animal advocates, they emphasize that vegetarianism is a personal choice, and that people should be free to decide for themselves how to eat. Consumers defend themselves with this argument, telling vegetarians things like, "I think it's great that you're vegetarian, but it's my personal choice to eat meat." Because individual messaging inescapably focuses on an individual's personal decision, it loads this counterargument into the minds of consumers, which stacks the deck against advocates. It could even lead to people thinking of veg eating as a trend or fad, further weakening the moral impetus behind it.[18]

One example of this counterargument playing out is the social media backlash the Good Food Institute received for its campaign to get In-N-Out Burger to carry a veggie burger. People on social media reacted negatively, drawing upon the personal-choice mindset with criticisms like "You don't see me protesting at vegan spots for not having meat options."[19]

While this seems obviously wrong to advocates, it's actually pretty reasonable from the perspective of someone who has only been exposed to individual messaging.

Social pressure

As discussed in the last chapter, there is abundant psychological evidence for the power of social pressure. While both individual and institutional messaging can incorporate social pressure, institutional messaging has more of it built in because it necessarily communicates that fighting animal farming is a group effort.

Social pressure is also known as a "descriptive norm," a type of "social norm" that represents what people think their social group does: What are their religions? What are their lifestyle choices? What are their political views? The other type is an "injunctive norm," what people think their social group should be: Is gender inequality seen as morally acceptable? Is recycling seen as a good practice? Effects on either type of norm seem instrumental in effective social change, and social change campaigns that target injunctive norms are seen as especially effective.[20] In addition to the institutional focus's increased impact on descriptive norms, it seems to have more of an effect on injunctive norms, such as if an entire school or workplace adopts a Meatless Monday policy where eating plant-based is a collective goal.

But isn't consumer action a clearer call to action?

Individual messaging has a straightforward call to action. Changing your diet is something you can do immediately with a fairly direct impact, while the call to action for institutional change is usually less self-evident and its impact is less direct. The clarity of the individual focus could make the outreach recipient more likely to act on that call to action because people who hear an institutional message might agree with the message but not fully realize that agreement with the message implies that they should take action now. For example, the institutional-message recipients could think they should just wait until better animal-free foods are developed and then follow the crowd. It might not immediately occur to people that they can help facilitate that development by contacting a company or government representative, joining a protest, working for or donating to animal advocacy organizations or animal-free food startups, sharing articles and other media, or having conversations about animal farming with their friends and family. These have to be added to the message in addition to its main institutional component.

More people leaving animals or all animal-based foods off their plates could also lead to substantial spillover benefits. As we mentioned before, there is some empirical evidence that eating animals leads people to think animals have less sophisticated mental capacities, likely due to the cognitive dissonance people experience when they consider that animals have rich mental lives and can't reconcile that with eating them.[21] The attitude shift caused by reducing that dissonance could lead to more activist involvement as well as sustained diet Short-term change also has a short feedback loop, meaning advocates can repeatedly measure their impact and modify their strategy based on the outcomes. For example, you could vary whether you ask people to "go vegan" or "go vegetarian" and then measure which call to action leads to more animals spared. This research-on-the-go is more challenging for longer-term outcomes, though we can use short-term proxies such as how someone's attitude changed immediately after seeing an institutional message, which might be a good predictor of whether their behavior will also change down the road.

Putting it into practice

It's possible to mix and match the two types of messaging, especially given time to explain the message in detail, as in a personal conversation. You can tell people something like, "Help end animal farming: go vegetarian!" Some mixes of the two types might capture the benefits of each while mitigating the downsides, though even a mixed message would still require a push in the institutional direction relative to the movement's status quo.

I think the need to focus more on institutional messaging and interventions is the most important underappreciated research finding in the movement. Of course, we should continue using individual, consumer-focused strategies to an extent, such as by including diet change in a list of actions people can take to support the cause and by boycotting food companies when they refuse to make positive changes.

Trigger events

I don't want to end a discussion of institutional change without highlighting the importance of "trigger events," a term legal scholars and political scientists use to refer to "those immediate factors in the political setting which provide the link between demands for action and public policy."[22] The most impactful trigger events can catalyze movements overnight, hurdle an issue into the spotlight, and compel policymakers to take action. Common examples are corporate disasters that highlight a lack of regulation, such as the Triangle Shirtwaist Factory fire of 1911 and the 1989 Exxon *Valdez* oil spill, which led to stricter safety standards for the garment and oil industries, respectively.[23] They are often tragedies, like the recent high-profile cases of black people killed by police officers in the US, which gave rise to the Black Lives Matter movement and ignited national debates on racism.

Trigger events can also be created by activists. The US environmental movement has done this particularly well, such as with the publication of Rachel Carson's *Silent Spring* in 1962 and the creation of the first Earth Day celebration in 1970.[24] The Salt March in India and Mahatma Gandhi's hunger strikes are well-known examples from anticolonial movements. The self-immolation of Mohamed Bouazizi in 2010 and the Tiananmen Square protests in 1989 fueled movements against oppressive governments in the Arab world and China, respectively. The arguments discussed in favor of institutional change also apply to these strategies.

Trigger events excel at communicating the urgency of the issue. They generate news, often because of some ticking time bomb situation, like the unregulated pesticide use spotlighted by *Silent Spring*. In France, the 1973 oil crisis—when oil was embargoed in the Middle East, leading to more than a three times' increase in price—was key to French adoption of nuclear energy. Because French support for nuclear was already gaining momentum and the country was so dependent on imported oil, this shock helped the pro-nuclear movement increase the nuclear proportion of energy generation from 6 percent in the early 1970s to 76.9 percent in 2014. The example of French nuclear energy also suggests that the farmed animal movement should focus on countries with an unreliable animal products supply, such that they could benefit more from food security. One candidate is Singapore, which imports roughly 90 percent of its food and has a history of ambitious centralized food and health technology programs.[25]

Small activism events, such as hosting local protests and social events, have immediate impact, such as helping people stick to vegetarian and vegan diets, but advocates should focus on the role of these events in the bigger picture of trigger events and the stepwise progression of social movements. This can help activists avoid common pitfalls, such as using PETA's offensive and trivializing strategies like "booth babes" who advertise animal-free food. Those tactics garner short-term attention at the cost of long-term credibility, reducing the ability to effect high-impact trigger events in the future.[26]

Examples of trigger events in the farmed animal movement include the 2008 undercover investigation by the Humane Society of the United States that led to the largest US meat recall in history; the showcase of the world's first cultured hamburger in 2013; the David and Goliath battle between Hampton Creek and the egg industry; and the 2017 film *Okja* that told a compelling fictional story about a genetically modified "super pig" rescued by animal rights activists. Each of these events signaled to the public, the animal agriculture industry, and the US government that the movement has significant capacity for future progress. We can sway public opinion. We can organize. We can produce concrete results in the social, political, and corporate realms. These signals are key to changing institutional behavior.

How over *why*

In order for an institution or individual to change their behavior in response to advocacy, they need to have both the motivation to make the change and the tools to do so. For example, with diet change, motivation can derive from ethics, taste preferences, price, and so on, and the primary tool people need is an understanding of how to find and prepare animal-free food. This matters whether we're focused on changing the diets on a one-by-one basis, as I've cautioned against, or changing the eating habits of society as a whole. Currently, it seems like more marginal resources should be spent on the *how* than on the *why*. This is mostly based on the following evidence.

AUDIENCE SELF-REPORT When advocates hand someone a leaflet on the street, show their friend a video of undercover investigations, or speak with a journalist about animal-free eating, the hesitation and counterarguments we hear are mostly about *how* they can change their behavior, not *why* they should. Common concerns include

- "I'm an athlete. Where would I get my protein?"
- "It's just so hard to find vegetarian options when eating out."
- "I would love to be vegan, but I could never give up cheese."

It's become increasingly less common over the past few years to hear arguments against changing to a non-meat diet such as

- "They're just animals. They don't matter."
- "Most farms aren't like the ones in that investigation."
- "I only buy meat from humane farms."

As discussed in chapter 6, the last of these counterarguments is still pervasive and is particularly obstructive, but overall, advocates hear many more of the *how* rebuttals. In fact, it's very common for an advocate to approach a conversation with the intention of persuading someone to go vegetarian, only to quickly realize that the person already agrees in principle that they should but just doesn't know how.

This finding has been demonstrated in a particularly quantifiable form of advocacy: the online ad. Mercy For Animals has tested both *why* ads that focus on the cruelty of animal farming and *how* ads that simply advertise the nonprofit's Vegetarian Starter Guide for viewers whose online activity suggested they're already interested in going vegetarian. The starter-guide ads cost less than a fifth as much per conversion, $0.20 compared to $1.10, with cost per conversion defined as the amount of advertising expenses it took to get one person to enter their email address, which was framed as a commitment to vegetarianism where the viewer would be sent a free starter guide.[27]

NEGLECTEDNESS As discussed in chapter 3, once people go vegan in an overwhelmingly meat-eating society, they can feel a desperate urge to shout from the rooftops about the cruelty and environmental devastation of animal farming. They see their friends and family as participating in a moral catastrophe, and assume that if those people just became educated about farmed animal abuses, they would surely take action. Unfortunately, for the reasons discussed in the last chapter, the problem is not as simple as that. Knowledge of harm alone is usually not enough to change us, even if we agree that what's happening is horrible.

Over the last couple of decades, much of the US and Europe has been exposed to the information on *why* animal farming is such an atrocity, but for much of that time advocates largely neglected the *how*. Advocates should look at current resource allocation to inform their use of marginal resources because the neglected strategy often involves more bang for your buck. Climate change advocates have made similar criticisms of their own movement, arguing they have focused too much on fear-mongering and too little on solution-focused framings.[28]

When people are provided with an achievable path to a better world, like with institutional messaging, people become not just more able to take short-term action but they become fundamentally more concerned. For this reason, a lack of *how* information might not only make it hard for people to act on *why* information, but could even be making people less receptive to the *why* information in the first place.

Stories, then statistics

This argument is a fairly uncontroversial one, but I find that advocates who fully agree with it still often don't fully apply it. When we discuss the harms of animal farming, we should focus on individual victims, sharing their stories as examples of the larger issues, instead of or before conveying that information with abstract facts and statistics. If advocates want to convey the vastness of the suffering in the animal agriculture industry, we need to first illustrate a single individual's experience, then work to help the audience understand that many more animals are in similarly dire situations.[29]

In psychology, the huge appeal of one affected individual is referred to as the "identifiable victim effect." The go-to example of this is the case of Jessica McClure, an eighteen-month-old infant who fell into a well in Midland, Texas, in 1987. She was rescued in fifty-eight hours by emergency personnel, and the family was inundated with gifts, cards, over $700,000 in cash, a visit from Vice President George H. W. Bush, and a phone call from President Ronald Reagan.[30] Meanwhile, many charities struggle for even a fraction of that attention to help many more children at risk of dying from other causes, particularly those in low-income countries who suffer from diseases like malaria.

Similar psychological effects include the collapse of compassion, discussed earlier, and scope insensitivity. The canonical example of the latter is an experiment where researchers told participants about the issue of waste-oil holding ponds created by the oil and gas industries. These pools cause the deaths of many migratory birds in the southern United States who land in them without realizing they're unsafe. Participants were divided into three groups and told that a new project could save two thousand, twenty thousand, or two hundred thousand of these birds. Researchers asked them how much their household would be willing to pay to help these animals. The average willingness to pay for each group was $80, $78, and $88, respectively. That means participants who were told ten or one hundred times as many birds could be saved were willing to contribute only about as much as participants told of the lesser amount.[31]

While many of us would prefer a world where these biases did not exist, they do, so advocates must take them into account when conveying the harms of important social issues. One caution with the story-based approach is to avoid going overboard with it and appearing as if animal advocates want to rescue every farmed animal one by one. Statistics are still important to include, especially with intellectual audiences, to clarify the magnitude and rational arguments for tackling this issue.

Be cautious with confrontation

Should we protest? Should we yell? Should we conduct sit-ins and other acts of civil disobedience to draw attention to the moral catastrophe of animal farming? This is one of the most polarizing and frequently debated topics in the movement. Some people feel these disruptive actions are effective tools for sparking conversations that lead the audience to see the harms of animal farming. Others worry that any benefit from this attention is outweighed by the impression the audience gets of advocates as angry fundamentalists and the defensiveness they feel when it seems like advocates are attacking them. Both these intuitions seem valid, and I don't think the aggregation of people's intuitions on the topic provides much evidence in either direction, so we need to look at the psychological and sociological evidence.[32]

Historical precedent

The main evidence besides intuition that advocates cite in favor of confrontation is its apparently important role in the most famous historical social movements. One of the earliest examples of this comes from the US antislavery movement; confrontation was correlated with major success after approximately 1807 when the US and the UK abolished the international slave trade.[33] Confrontation played a key role in human rights movements of the mid-twentieth century, and because of its eye-catching nature, confrontational strategies like sit-ins and marches are often the first things that come to mind when one thinks about activism.

However, these examples come with important qualifications. Successful confrontation often happens once public support is already substantial, the topic is already mainstream, and significant policy victories have already occurred. Arguably, the farmed animal movement is not as far along as oft-cited movements like post-1807 antislavery, so confrontation might be less effective.

Emotional arousal and moral outrage

As with institutional messaging, one of the key components of confrontational activism is how it inspires emotional arousal and moral outrage. If you have ever participated in a protest or even watched a video of one online, you probably experienced some sort of heightened emotion, much more than occurs when you hand out or receive a leaflet.

The main benefits of emotional arousal and moral outrage are making content viral; helping the audience overcome status quo bias and system justification; and increasing the likelihood of the audience not just changing their personal behavior, but becoming advocates and multiplying their impact.[34]

But provocative nonconfrontational tactics like undercover investigations can have all of these effects, too, without as much risk of potentially backfiring, moving some of the audience away from the advocate's position—though note that the backfire effect is an active field of research, and the original study documenting it has so far failed to be replicated.[35] Of course, confrontational advocacy can take many forms, but I worry that advocates currently engaging in confrontation gravitate toward less promising tactics, such as throwing fake blood or dressing in animal costumes. While these tactics get attention, and that makes them popular with advocates, pure attention doesn't seem to be worth the cost of making animal-free food advocacy seem juvenile, rude, trivial, laughable, or otherwise not fitting of the moral consideration given to more mainstream social justice issues such as civil rights.

As described in chapter 2, the creation of this negative public attitude—or at least its exacerbation—seems like one of the worst mistakes of the movement to date along with the heavy focus on consumer change. Instead of these tactics, advocates can take part in marches, in peaceful protests that are professional and respectable, and in acts of civil disobedience that don't involve yelling at consumers or other forms of aggression, such as the temporary blocking of slaughterhouse trucks from reaching their destination, a tactic that often involves offering the dehydrated animals water, allowing them to experience at least one compassionate action in their lives.

Another best-of-both-worlds strategy is to make our rhetoric kind and accepting, but also bold. Advocates should be bold and radical by honestly explaining the huge scale and horror of animal farming, but we should also be considerate and accepting of our audience. That means not blaming them for the problem, and not treating them badly simply because they have yet to get on board with the movement. For example, in a protest at a restaurant, it's important to maintain distance from the consumers, such as by standing outside the restaurant, and to focus rhetoric on the restaurant company, not on the consumers therein. This balanced approach can reap much of the benefit of moral outrage while mitigating the defensiveness seen in response to more aggressive confrontational activism. We should be radical in our message, not necessarily in our tactics.

8. BROADENING HORIZONS

IN THE nineteenth and early twentieth centuries, vegetarianism was a diet of self-deprivation. It was largely chosen by people who wanted to deny themselves the pleasure of eating animal flesh. That's a fine personal choice, of course, but it appeals to only a very small subset of the population, especially today when the taste and experience of food is seen as one of life's greatest pleasures. Today, animal-free food is broadening far beyond the ascetics, which is fantastic, but the expansion requires a careful look at existing social dynamics like race, gender, political affiliation, and geopolitical identity. These broadening horizons are the final leg of our strategic roadmap to an animal-free food system.

Not just for hippies

WHAT DOES A VEGETARIAN LOOK LIKE? Most advocacy to date has been based in North America, Europe, Australia, and New Zealand, though there's frequent incorporation of Asian cuisine and appeals to Asian religions. Within those high-income Western countries, the movement has been associated with specific demographics: white, female, young, well-educated, wealthy, idealistic, and liberal. This association can unfortunately lead people of other demographics to feel uncomfortable with vegetarianism and animal-free food. It severely limits the movement's ability to achieve its goals.

But what is the actual composition of the farmed animal movement?

We have some survey data for US vegetarians. A 2006 Vegetarian Resource Group nationally representative US poll found that 5 percent of males self-identified as vegetarian compared to 9 percent of females, and 6 percent of white respondents reported being vegetarian along with 7 percent of black respondents and 8 percent of Hispanic respondents. The margin of error here is the standard +/- 3 percent, meaning this is only very weak evidence of differences between ethnic groups. A 2012 Gallup poll found 4 percent in males and 7 percent in females, and it found that vegetarianism rates were higher in people age fifty or older, people who didn't graduate from college, liberals, and unmarried individuals. All of these differences were within the 4 percent margin of error, and Gallup didn't publish race information but did weight samples by race and stated that "almost all segments of the U.S. population have similar percentages of vegetarians." In 2015, the nonprofit organization Faunalytics conducted a poll with similar results.[1]

If we look at the amount of meat consumed in the US in 2012—the most recent data available—we see that people who self-identify as white, Hispanic, and "other" consume around the same amount, at 191.2, 193.2, and 191.5 pounds per year respectively. On average, black people consume a higher amount at 236.0 pounds and have the highest consumption of fish, turkeys, and chickens.[2] Given that these animals suffer more per pound due to their small size and exceptionally cruel treatment, black advocates have called for an increased focus on outreach to black communities, both to address these animal welfare concerns and to mitigate public health epidemics. Many have attributed the health issues to America's history of racism because unhealthy food has often been all that was available and affordable to black people.[3]

Overall, it seems that vegetarian stereotypes only partially match reality, but reality is only one part of the issue. As writer Aph Ko notes, "Veganism isn't *really* white, but the media coverage of it *is*." She demonstrates this with a clip from the popular TV show *Orange Is the New Black*, where a group of black women in prison make fun of veganism, yoga, and wine tasting as "white people politics." Ko goes on to say that some activists have perpetuated this stereotype, excluded black people from the animal rights movement, and facilitated racism by making insensitive comparisons between the mistreatment of black people and the mistreatment of animals.[4]

But this leads to another question: How did vegetarianism get associated with whiteness if this doesn't reflect its reality? A chief suspect is the heavy consumer focus of the movement to date. As discussed in chapter 7, advocates have strongly associated the issues of animal farming with a call for individual consumption changes. Almost every video released has a call to action like "leave animals off your plate." This has led vegetarian and vegan foods to be seen as high-end fare similar to organic, free-trade, and local foods. All of these labels are commonly seen as catering to upper-class, white, and liberal consumers, who are seen as willing to pay a premium for personal purity and warm, fuzzy feelings.

This association with luxury is especially unfortunate given that animal-free foods actually tend to be cheaper than animal-based foods, and far cheaper if we consider the high costs of animal farming to the environment and economy. It's simply not cost-effective to feed an animal ten or more calories of plants so she can grow one calorie of meat.[5] Indeed, popular vegan foods that lack explicit vegan labels, such as pasta, beans, rice, lentils, and peanut butter, are inexpensive staples.

Inclusive advocacy

People who see vegetarianism and animal-free food advocacy as outside their demographic are probably more hesitant to change their diet or join the movement. I had some personal experience with those stereotypes in Texas: most of my friends were politically right-wing and they quickly associated my vegetarianism with hippies, Democrats, and sentimentality. Vegetarians will often ask meat eaters questions like, "You wouldn't kill an animal with your own hands, so why pay someone else to do it for you?" expecting this to be a persuasive argument, but many people I grew up with were perfectly happy to kill animals themselves, whether animals they raised on small noncommercial farms or wild animals they hunted. In fact, they thought, perhaps correctly, that this was more ethical than purchasing meat from factory farms.

This doesn't mean that there aren't compelling reasons for rural Texans to go vegetarian, just that the current popular arguments and framings are not well suited to their morals and motivations. According to the "moral foundations theory" proposed by psychologists Jonathan Haidt and Jesse Graham and popularized in Haidt's book *The Righteous Mind*, human morality is based on six foundations:

- Care/Harm, such as caring for suffering animals
- Liberty/Oppression, such as opposing bullies or supporting affirmative action
- Fairness/Cheating, such as wanting high-performers to be rewarded in the workplace
- Loyalty/Betrayal, such as punishing someone who commits treason
- Authority/Subversion, such as respecting police and elders in a community
- Sanctity/Degradation, such as avoiding sexual promiscuity or taboo ideas like communism in the US

Research suggests that US liberal voters and politicians focus on Care/Harm, Liberty/Oppression, and Fairness/Cheating, though they are willing to give up Fairness for the sake of the other two. On the other hand, conservatives utilize and appeal to all six of the foundations. This can lead to challenges for liberals understanding and communicating with conservatives. Often liberals will look at a problem solely in terms of harm and oppression, judge a policy choice on only those criteria, and think that conservatives have to be crazy or lying to take a different position. So if liberals want to appeal to conservatives, for instance to convince them to oppose animal farming, they should try to consider the other foundations and discuss them in their arguments as well.[6]

Advocates should also avoid focusing too much on terms like "oppression" and "compassion" when appealing to conservative or bipartisan audiences. These not only fail to accommodate the last three foundations; they signal that the speaker is strongly liberal, meaning they're an outsider with less authority from the perspective of a conservative listener. Scientists have seen this sort of effect with climate change advocacy. They found that portraying climate change action as a way of uniting society around a common goal, or as a way of spurring economic and technological development, was more effective in reaching conservatives than the usual emphasis on avoiding ecological disaster.[7]

One other strategy for reaching conservatives is to emphasize how animal-free food can preserve the parts of the status quo that conservatives care about. This strategy has also worked for climate change. Researchers found that US conservatives could be persuaded to care about climate change with a message like, "Being pro-environmental allows us to protect and preserve the American way of life. It is patriotic to conserve the country's natural resources." Another study found that the so-called "vegetarian threat," meaning agreement with statements like "The rise of vegetarianism poses a threat to our country's cultural customs," was an important mediator of conservative opposition to vegetarianism.[8] So if animal-free food can maintain these customs, such as the tradition of barbecue in the southern US, it could reach a more bipartisan audience.

Patriotic and national security messages have been highly useful in historical social movements. For example, in *Brown v. Board of Education*, the Truman administration filed a court brief calling for the integration of schools largely based on American success in the Cold War. It said: "The United States is trying to prove to the people of the world of every nationality, race and color, that a free democracy is the most civilized and most secure form of government yet devised by man. . . . The existence of discrimination against minority groups in the United States has an adverse effect upon our relations with other countries. Racial discrimination furnishes grist for the Communist propaganda mills."[9]

Farmed animal advocates can similarly frame fixing our food system as a point of national pride. For example, we identify as a nation of animal lovers, yet so many animals suffer on US factory farms. We're a nation of innovators, so we should take the lead on the revolutionary new animal-free food technologies.

To broaden the movement's horizons, advocates should also ensure that imagery, such as on billboards and in leaflets, avoids vegetarian stereotypes. In addition to featuring racial and gender diversity, advocates can showcase the many world-class vegan athletes who maintain the strength and power required in their profession on an animal-free diet. This clearly communicates that physical strength is not an issue with well-planned animal-free diets. Indeed, many argue that it increases performance.[10]

Similarly, advocates can emphasize the affordability of animal-free foods and their benefits to low-income consumers. Toni Okamoto is leading the charge with her content platform, Plant Based on a Budget. She shows the public how "affordable, easy, and delicious a plant based diet can be." Given how huge a driver price is of consumer choices, even in high-income countries like the US, this project fills an important strategic gap in the movement.[11]

The movement still has a long way to go to shed stereotypes and increase inclusivity. It's not immune to the gender inequalities that operate in society. Even though most vegetarians and most animal advocates are women, leaders of animal-free food companies and advocacy nonprofits are still largely men. Farmed animal advocates can't expect to end all oppression and discrimination in society by themselves, but the movement could participate in the broader push for racial and gender equity, both for the benefit of marginalized groups and for the movement's own success.

The following are some ways the movement can broaden its horizons within high-income Western countries. While these suggestions are some of the least controversial in this book, putting them into practice is no small challenge:

- Account for different empirical and moral perspectives when crafting a message, such as considering all six moral foundations when speaking with conservatives.
- Represent all demographics of the movement, especially via leadership and public roles.
- Avoid messages that trivialize and/or undercut other social movements, such as shaming overweight people in order to communicate the health benefits of animal-free food.
- Learn about important social issues besides animal farming, such as racial and class discrimination, and the intersections of these issues. The more informed

advocates are, the better prepared we are to address issues of diversity and inclusion.

- Try to build bridges with other movements. If you have time and interest, join them for their own sake, helping with activism, funding, or in other ways.
- While it may be tempting to focus our outreach on the most receptive demographics (often thought of as young women), it's important to consider the downside of creating a homogeneous movement, narrowing our horizons, and exacerbating stereotypes.[12]
- Carefully listen to minority opinions within the movement.
- Avoid inconsistent messages about animal-free food when possible, such as describing it as delicious while pointing to expensive processed products and simultaneously claiming it's affordable based on the cost of whole grains, beans, and vegetables. Of course, you can talk about the higher-end foods available as well as the low cost of other items, but try to avoid misrepresenting the category as a whole.
- Encourage other advocates to use these strategies, and to think more critically about this topic. We should continue to consider whether these strategies actually are the most promising, because the research is always evolving.

Reaching around the globe

As the movement expands around the globe, advocates need to make tough decisions about which regions to prioritize and how to adapt their messaging and strategies to different cultures. In this section, we'll consider which countries to prioritize working in and how to go about that expansion.

WHERE ARE THE MOST ANIMALS SUFFERING? The Open Philanthropy Project tackles the world's most pressing problems. It's a collaboration between GiveWell, one of the first organizations in the effective altruism movement and now the world's leading impact-focused charity evaluator, and Good Ventures, the $8 billion foundation established by Facebook cofounder Dustin Moskovitz and his wife, Cari Tuna. They're now the leading funder in the farmed animal movement and, like other effective altruist organizations, are tackling the challenge of international expansion. To do this, Lewis Bollard, an advocacy researcher at the Open Philanthropy Project, first asked how many animals live on factory farms at any given time in different countries. He compiled data from the United Nations Food and Agriculture Organization and Fish Count, which is recognized as the leading organization tracking the number of fish killed and consumed globally.

What Bollard found is that approximately 49 percent of all farmed animals live in a single country: China. There's significant uncertainty in this figure of sixty billion animals, but it's clearly much higher than the next largest: eight billion in India. Moreover, as the quickly growing populations of these countries become more wealthy, they increase their consumption of animal products, which are more expensive.[13]

Pei Su is a Taiwanese activist leading animal protection campaigns in East Asia. Su's father was Chinese, but he left mainland China for Taiwan after Chiang Kai-shek's defeat by Mao Zedong and the Chinese Communist Party in 1949. Su grew up in the White Terror, a thirty-eight-year period of anti-communist authoritarian rule in Taiwan. There was a curfew for all citizens between midnight and 5 a.m. In school, students were required to speak Mandarin instead of the native dialect of Taiwanese. Everyone lived cautiously, avoiding any actions or statements that could be interpreted as sympathetic to the Communist Party.

Taiwan was poor. Meat was rare. In East Asia, it was usually considered a delicacy or garnish on a meal of vegetables, fruit, or soy foods like tofu. The Chinese character for *plain*, 素 (pronounced "sù"), is actually the same character used to refer to vegetarian food. Less than a handful of times each year, her family got to enjoy a whole chicken. The most desirable part was the legs, which usually went to Su's father and brothers. On her birthday, Su was allowed a chicken egg or drumstick. When I spoke with her, she referred to big meat portions like steaks as a "Western concept," a trend that has been quickly growing in popularity in East Asia over the last half-century. Dairy was introduced to the region while Su was still in school and promoted to families as a healthy, special treat that signaled their growing prosperity.

Su's mother, father, grandmother, and aunt all passed away when she was a young teenager. She learned about the regular beatings of peaceful protesters and the Tiananmen Square Massacre in 1989. She told me, "The Tiananmen Square Massacre filled me with an overwhelming sense of injustice—death and killing made no sense to me." This inspired her to become a Buddhist and vegetarian. She became an activist when Taiwan transitioned to democratic rule.

Su eventually quit her day job of running a flower shop and started working full-time for the only Taiwanese organization actively working to draft animal protection legislation. She and a Buddhist monk conducted undercover investigations of animal farms and slaughterhouses, sharing the information with domestic and international contacts. Su realized how compelling animal protection was for Westerners, so she decided to travel to the US and Europe to learn how organizations there had been so successful. She completed a master's degree in the UK in social policy and animal rights, and now works with international and local East Asian groups to effect change for animals with her nonprofit ACTAsia. She and other advocates agree that the best groups to work directly on animal protection in a particular region are local groups whose members have a much better understanding of and ability to work with local culture and institutions.

Unfortunately, China's animal protection movement still lags behind similar movements in the West. Taiwan has made some progress, but mainland China had no animal protection laws at the time of this writing. China has also rapidly adopted a factory-farming food system. In 2000, around 35 percent of chickens raised for meat came from farms with more than two thousand birds. By 2009, the figure had risen to 70 percent.[14] Advocacy in China currently focuses mostly on simply raising awareness of animal issues. In 2014, Su's organization ACTAsia hosted China's first animal-free fashion show, which featured no fur or animal-tested makeup and served entirely vegan fare.

While animal welfare discussions are less common in China than in the US and Europe, a 2011 Chinese survey did find that 70 percent of respondents thought it was "somewhat inappropriate" or "extremely inappropriate" to use a cement floor on pig farms, and 74 percent felt similarly about killing fowl near cages that housed live fowl. In the US, polls have suggested around 80 percent oppose similar blatantly cruel practices like fast-growth chicken breeding and extreme crowding. In that same Chinese survey, 72 percent expressed concern about "overusing additives," 50 percent were concerned about "overusing antibiotics," and 48 percent were concerned about "bad taste" caused by factory farming practices. However, significant minorities praised factory farming for "high profits" (22 percent), "high productivity" (33 percent), and "fast growth for slaughter" (38 percent). There haven't been any similar questions put to Americans. Finally, only 37 percent of respondents had heard of the Chinese term for "animal welfare," which, as an American, is hard for me to imagine. When I first heard that low percentage, I hardly believed it, but Chinese advocates I've spoken with have confirmed that "animal welfare" is just not an established concept in their country.[15]

What about India, the country with the second-largest number of farmed animals? Bollard describes India as "a nation of contradictions" for animals: it has the world's largest number of vegetarians despite being the second-largest consumer of animals. This is partly due to India moving away from red meat and toward poultry and eggs, similar to the rise in poultry consumption we saw in the US starting in the second half of the twentieth century, though US egg consumption per capita has decreased while India consumption has increased.[16]

Some advocates are less concerned about animal farming in low-income countries like India because they presume that farms are more pastoral, natural, and humane. We already saw that China is moving animals into factories at a rapid pace, and unfortunately India has already adopted the factory farming system as its dominant method of animal farming. As Bollard learned in a 2017 research visit to India, even the animals' genes are sometimes from US companies, meaning the animals there are also bred to suffer from overactive meat, dairy, or egg production.

When I spoke with him, Bollard told me about the numerous challenges of working in India. One is that the retail system is distributed and diverse. For example, 90 percent of chickens raised for meat are sold live at local markets, so advocates can't simply target a few large retailers like they can in the West. Additionally, there is a substantial controversy in India over the consumption of cow meat—cows are considered sacred by some Hindus, so advocacy on behalf of cows can be perceived as nationalist and intolerant of non-Hindu religions. Animal advocates try to steer clear of this, such as by focusing on chickens.

The last thing to consider about India is that it has fairly strong animal welfare laws. For instance, in 1960, the Indian parliament passed a remarkably progressive law to prohibit the confinement of animals in cages without "a reasonable opportunity for movement," and in 2012, the government warned the egg industry that battery cages didn't meet this criterion and needed to be phased out by 2017. However, every farm Bollard visited in his 2017 trip to India still used battery cages and none had plans to transition. One explanation for this discrepancy is that the laws simply lack meaningful enforcement. Even if charges are brought against a farm with battery cages, the maximum penalty is only 100 rupees, which is the equivalent of about $5.80. This suggests that a different, enforcement-focused advocacy approach is needed in India.[17]

CHANGING THE WORLD Bollard lists his Big Four priority regions for the farmed animal advocacy movement as China, the EU, the US, and India. Even though the EU and US have only around a billion farmed animals each, they join China and India in this list because of their significant influence over the global landscape and because of the existing momentum for change in these regions.[18]

Every year, *US News & World Report* collaborates with BAV Consulting and the Wharton School of the University of Pennsylvania to conduct a global survey asking for participants' views of different countries. Number one on the list for international influence in 2016 and 2017 was the US, which, the agency speculates, is due to its military power, the popularity of its media and entertainment industry, and its position as the world's largest economy. The UK, Germany, and China were in the top five both years, along with France in 2016 and Russia in 2017, perhaps due to the involvement of Russia in international affairs like the war in Syria. This suggests that new food trends and policies in these countries could have a disproportionate influence on other countries, even if fewer animals are directly affected.[19]

Unfortunately, it may be harder to make progress in the US due to the heavy influence of corporations, especially the animal agriculture industry, on society and government.[20] The US leads the world in meat consumption per capita and has developed strong associations between animal products and cultural identity. It also lags behind many other high-income countries in terms of its policies on social issues relevant to animal farming: It's ranked twenty-sixth in the 2016 Environmental Performance Index, the most reputable such metric. It earned a D rating in animal welfare standards from the international nonprofit World Animal Protection, though Canada and Japan also received Ds. And the US health-care system is less efficient than that of most other high-income countries, given the extraordinarily high rates of obesity and diet-related diseases in the US.[21]

The tractability of advocacy in China and India is more uncertain. China's rapidly industrializing economy is causing more damage to the environment annually than that of any other country, but it has become a global leader in environmental policy.[22] Chinese consumers eat around 173 grams of meat per day, but the government recommends only 40 to 75 grams—less than an average American hamburger patty.[23] China has a highly centralized governance system, which makes policy change more difficult, but also makes changes easier to promulgate across the country. Meat has been regarded as a luxury, but it also hasn't been as associated with Chinese cultural identity the same way bacon, cheese, and bratwurst have in many American and European cultures.

International organizations have had some success encouraging Chinese companies to reduce farmed animal suffering. For example, the UK-based nonprofit Compassion in World Farming gave their Good Pig Production awards to Chinese food companies with progressive welfare policies.[24] The Open Philanthropy Project has granted millions of dollars to nonprofits for more of these programs. However, Chinese government regulations of foreign charity involvement are a significant barrier to international organizations, yet another reason to work with and focus on creating and supporting local groups.[25] International companies are also applying pressure: McDonald's sent animal scientist Temple Grandin to China to conduct animal welfare training courses and has increased audits in response to food-safety scandals.[26] Bollard thinks that the numerous food-safety scandals in China, such as the milk dilution discussed in chapter 4, have been a key force for the Chinese government and corporations to address factory farming issues.

Hong Kong is already seeing promising signs of a transition toward animal-free food. The estimated number of residents who are vegetarian at least one day a week has increased from 5 percent in 2008 to 23 percent in 2014. The number of vegetarian restaurants increased from 140 in 2013 to 240 in 2016.[27] This is partly attributable to the success of Green Monday, a company founded by two vegetarians that promotes healthy, sustainable diets and lifestyles. Its leader, David Yeung, helped make Hong Kong the first city outside the US to sell the Beyond Burger.

Yeung told me that he's quite optimistic about the future of animal-free food in China. In addition to the concerns about food safety that Bollard cited, Yeung thinks animal-free foods have a special appeal in East Asia because of vegetarianism in popular religious traditions. Yeung's father was a businessman, committed philanthropist, and devout Buddhist who raised him with traditions like eating vegetarian on the first and fifteenth day of each lunar calendar month. Yeung left Hong Kong to study engineering at Columbia University in New York. While in the US, he read more on Buddhist philosophy and went vegetarian, coming to believe that the ethical idea of karma made sense in the same way as Newton's third law of motion: Every action has an equal and opposite reaction.

In recent decades, many young people in China have become distanced from Chinese philosophical traditions.[28] However, Yeung sees a trend back toward them due to both deeply rooted cultural norms and the alignment of some traditions with modern Western interests in sustainability, personal wellness, and animal welfare.

I agree with Bollard that these four regions—China, the US, the EU, and India, in that order—seem to be the most important for the farmed animal movement. However, these nations hold only around 60 percent of the world's farmed animals.[29] Elsewhere in the world, there are scattered animal-free food companies popping up such as the Not Company in Chile with products such as NotMayo and NotCheese. NotCo, as the company refers to itself, uses machine learning to combine diverse plant ingredients and create the right culinary experience for its customers.[30] We also have highly effective activism strategies being replicated across the world, such as Animal Equality running undercover investigations in Latin America, as it has already successfully done in Europe.[31]

We also shouldn't forget the companies that produce the traditional animal-free foods like tofu, which are especially popular in East Asia and likely to remain more accessible than high-tech foods like cultured meat, just as they've historically been more accessible than animal protein.

I spoke to Prince Khaled bin Alwaleed bin Talal of Saudi Arabia about the situation in the Middle East and North Africa, a region known for its popular plant-based dishes like hummus and falafel. Bin Alwaleed grew up in a $136-million, 460,000 square foot palace and as a young adult took pride in his collection of around two hundred luxury cars. Today he owns only one, an electric Tesla. As an accomplished businessman following in the footsteps of his father, Prince Al-Waleed Bin Talal bin Abdulaziz al Saud—one of the world's richest men with a 2017 net worth of $17 billion—he has the resources and skills to lead the movement in this politically and economically important region.[32]

Bin Alwaleed was born in California and has spent much of his life in the West. He told me that Middle Eastern advocates should remember not to reinvent the wheel. They can simply bring the products of companies like Hampton Creek and Miyoko's Kitchen to their supermarkets, instead of forming new companies and products. In terms of public perception, the situation is similar to that of

China: people lack the foundational concern for animal cruelty that we can take for granted in the US. Bin Alwaleed sees documentaries as a tried-and-true tactic of the West that could be delivered to the Middle Eastern masses via Arabic subtitles. Like in other regions, plant-based eating is currently stereotyped in the Middle East as a niche lifestyle choice for yoga devotees and other small populations. This is why Bin Alwaleed prioritizes mass outreach, especially through social media, to show the public that everyone needs to shift away from animal farming. Similarly, lowering the price points of plant-based meat, dairy, and eggs could make a big difference.

The movement is expanding to all corners of the globe through its use of welfare reforms, such as cage-free campaigns for egg-laying hens in neglected regions like Eastern Europe, Latin America, and Africa. This work has been largely coordinated by the Open Wing Alliance, a collaboration spearheaded by The Humane League. While this work has obvious direct benefits for animals, it also inspires the creation and growth of local nonprofits focused on the issues of animal farming. This is leading to a global infrastructure for the production and popularization of animal-free foods. It could be a long time before countries that haven't even established a single law for farmed animal treatment start considering whether to get rid of animal farming altogether, but increasing globalization suggests that might happen sooner than we would expect. Animal farming is a global issue with wide-reaching consequences, and impact-focused advocates need to remember that big picture and avoid focusing too narrowly on the change happening in our local communities.

To achieve global success, the movement should prioritize work in China, the US, the EU, and India. Individual advocates should keep in mind that personal fit such as cultural familiarity, citizenship, and language fluency are very important for deciding where to work. It's also important to avoid singling out regions or cultures for reasons other than the total scale of harm. We've already seen significant harm come from the heavy Western criticism of China for the quite uncommon practice of eating dogs, despite the far greater suffering involved in the farming of chickens and fish. Not only can this perpetuate racism, but it can discourage other regions from identifying at all with the farmed animal movement. Why would you want to fight for a cause when its advocates are singling out your country as the enemy?

In terms of global organization, the movement should develop and support international organizations and networks that can be more effective by building bridges, sharing information, and coordinating campaigns. But when it comes to the actual day-to-day work of campaigning in foreign countries, advocates should hire, support, and listen to local advocates who probably better understand and communicate with their own culture.

9. THE EXPANDING MORAL CIRCLE, REVISITED

THE EXPANDING MORAL CIRCLE, fought for by advocates in every era of human history, could be the most important long-term project of the human species. The initial expansion from the individual to the tribe allowed us to successfully survive and flourish as hunter-gatherers. Extending it further allowed us to work together as agricultural communities, then as cities and countries. Today, we're still figuring out how to extend it to a global scale that allows for peace and international cooperation. We've also made significant strides, and have a ways to go, with fully expanding moral consideration to all genders, sexualities, ethnicities, and other demographics.

If our moral circle had expanded more fully and more quickly, we might have prevented or more quickly ended many of history's greatest atrocities, such as slavery, genocides, and wars. Atrocities rely on the exclusion of certain sentient beings from our moral consideration. We look back and praise the advocates who expanded the moral circle, but what will future historians say about the end of animal farming in this broader story?

Today, for the vast majority of humans—even most human rights advocates—the over one hundred billion animals suffering on farms and in slaughterhouses still lie at the edge of the moral circle.[1] This means that you, as a reader of this book, stand at the moral frontier. You've taken a huge step in considering the interests of farmed animals.

However, before I conclude this book, I want to ask: Are we who have opened our eyes to the harms of animal farming still ignorant of other classes of victims, perhaps who suffer in even greater numbers? Given that previous generations have been largely ignorant of the pressing issues we recognize today, it seems presumptuous to assume that we're the ones who have it all figured out. Yes, we've been around longer, but people in 1950 could have said that about people in 1850, and they still would have missed the major civil rights progress of the latter half of the 1900s. If we think most people in most other times and places have missed something important in their ideologies, how likely is it that our ideology happens to be the exception?

Looking forward

> *The mark of a civilized person is the ability to look at a column of numbers and weep.*
>
> —Unclear origin, probably adapted from Bertrand Russell

To discover our own moral blindspots, we can consider the dimensions by which our moral circle has already expanded to significant degrees—such as gender, ethnicity, location (from tribes to countries to the entire world), and species—and ask ourselves which similar dimensions exist upon which our moral circle has not yet fully expanded.

I'll take a look at a few of these excluded dimensions. Before that, I should note that considering these new frontiers of moral circle inclusion is challenging and can even feel like science fiction. But I'm not hoping to change your mind on these frontiers—I'm only asking you to keep your mind open.

It's important to consider that the modern issues of industrial farming would have also sounded absurd to past humans. For example, you might be familiar with *A Vindication of the Rights of Women*, one of the leading texts of feminism written in 1792 by Mary Wollstonecraft. She argued that girls deserved to have a formal education like boys had and that women should be able to openly speak their mind like men could. That same year, the philosopher Thomas Taylor responded to Wollstonecraft with *A Vindication of the Rights of Brutes*, where he argued that by the logic used to give women rights, animals should have rights too, which for him was clearly a reductio ad absurdum. Because animals clearly couldn't have rights, women couldn't either. Fortunately, history proved Taylor wrong.

Future humans and animals

The first dimension we'll look at is time: whether a being exists today, tomorrow, or even a thousand years from now. We regularly consider beings who haven't been born yet, such as our future grandchildren or the future generations who will bear the consequences of climate change or a loss of natural resources. But this is only the tip of the iceberg when it comes to future beings, who could exist in truly astronomical numbers.

Estimates suggest that there will be between 9.3 and 12.3 billion humans on Earth in 2100.[2] However, technological advances could enable future humans to not only cover our home planet but also colonize outer space, and might even be able to do so at an exponential rate as Earth-based space colonies then send out their own interplanetary colonists. This expansion could even start in the next few decades: tech mogul Elon Musk is hoping to send humans to Mars by 2025.[3]

So just how many people could exist after this rapid expansion? Researchers have estimated that in the long run there could be 10^{38} humans (and even more animals) if humanity colonizes the Virgo Supercluster, the massive concentration of galaxies that includes our own Milky Way galaxy and forty-seven thousand of its neighbors.[4] Interstellar expansion presents a tremendous opportunity for a progressive society to expand and flourish, but it's also a terrifying risk for the expansion of inequity, persecution, slavery, war, torture, genocide, and every other tragedy that's happened on Earth.

While it might seem speculative and unrealistic to consider a long-term and perhaps unlikely outcome like interstellar colonization, especially given all the suffering happening in our own backyard right now because of issues such as animal farming, the stakes involved are, quite literally, astronomical, so even a very small chance of affecting these extraordinarily important long-term outcomes could be a moral priority if we really want to do as much good as possible. (Proponents of prioritizing the "far future" argue that it doesn't depend on this Pascalian logic because we in fact have a pretty good chance of making a big difference in far future outcomes.)

One technology that could have a critical impact on the well-being of humans and animals in the far future is artificial intelligence (AI), so one way we could have an impact on the far future is through AI safety, working to ensure that AI has positive rather than negative effects on the world. One troubling scenario is if AI progresses slowly toward human-level intelligence, but then due to its ability to quickly improve upon itself, suddenly overtakes even the smartest human minds. Evolutionary processes took billions of years to shape modern biological intelligences, but a sufficiently advanced technology could modify itself, test those modifications, and learn to improve itself as dramatically in mere moments. It might be tempting to assume that humans would have total control over the AI's goals or to dismiss negative outcomes as science fiction, but experts in the field see value alignment—whether or not AI will have the same values as humans—as a very tricky problem, especially given factors like the competition between companies and countries to be the first ones to develop such a superintelligence.

There are only a few research teams working on AI safety relative to the many researchers working to build more powerful AI. Critical decisions might end up being made by only a handful of people, so someone with the right connections and information could have a significant impact on how AI advances.

Could it be the case that impact-focused do-gooders should deprioritize the suffering of humans and animals today, instead choosing to work on projects like AI safety that help beings who might or might not even come to exist hundreds or thousands of years from now? This argument might sound ridiculous, but farmed animal advocates should reflect on their own experiences with naysayers who assert, "It's crazy to think about chickens and fish when so many humans are suffering" or, "Your diet doesn't even make a difference in the scope of the global food system." Would we be making the same sort of error if we insisted on focusing only on those suffering now, even when we can impact so many others through work like AI safety? Doesn't it seem wrong to dismiss those countless future beings because our impact may feel like a drop in the bucket, even if in absolute terms it's enormous?

Consider the effective altruism prioritization framework of scale, neglectedness, and tractability. The far future has a scale that's literally astronomically greater than that of the near future (potentially 10^{38} humans at a time for millions of years!). It is also much more neglected: only a few dozen people are employed full-time on these projects. On the other hand, the far future does seem less tractable because, unlike near-term efforts, we don't have many feedback loops to see if what we're doing is working or much historical evidence to improve our effectiveness. We're left with the question: Is it astronomically less tractable, enough to outweigh its astronomically greater scale and its much greater neglectedness? Many effective altruists think not.[5]

Wild animals

The most populous group of currently excluded individuals, who may exist in astronomical numbers in the future as well, is the wild animals who exist outside of human civilization, in savannas, rain forests, lakes, oceans, and elsewhere. Many of us care about wild animals to some extent today, just like we care to some extent about future humans. We empathize with charismatic megafauna like pandas, elephants, and orcas. These species serve as spokespeople for modern conservation efforts. We sometimes care about individual, less charismatic animals, such as a wild sheep whose curved horn was stuck around a small tree. A jogger came across the distressed sheep in the woods and helped disentangle his horn, freeing him from what might have been a slow death from thirst. Video of the interaction was posted on YouTube, and viewers applauded the jogger for his courageous action.[6] The rescue of individual wild animals, from both natural and human-caused misfortune, is widely praised because we recognize that those animals have interests that matter, just like those of our pets. They have the capacity to feel suffering and happiness just like dogs, cats, and humans.

However, there are vastly more wild animals out there who suffer with no hope of rescue. They endure injury, illness, and starvation with astonishing frequency. Yet there has so far been very little research into or advocacy for large-scale interventions to improve their welfare, despite extremely large-scale impacts of humanity on their welfare through transportation, agriculture, and building construction. It's not a question of whether we should intervene in the wild, but whether we should continue with our current haphazard approach.

If we applaud the jogger who helped the sheep whose horn was wrapped around a tree or the wildlife rehabilitators who nurse individual animals back to good health, should we then also take systemic steps to prevent the suffering of *all* wild animals? One man in Kenya, Patrick Kilonzo Mwalua, is doing this work at the scale of his local community. The pea farmer drives a water truck around the savanna during severe droughts, filling empty watering holes for elephants, buffalo, and other animals who could suffer and die of thirst without his assistance.[7]

To be clear, what we're considering here is more than just conservation of natural habitat. Instead, it's the idea of actually intervening in nature to improve the welfare and protect the autonomy of individual wild animals who suffer intensely and in vast numbers.

When we assess possible interventions, we should keep in mind the "appeal to nature" fallacy, when someone argues that "nature is inherently good and therefore any intervention is necessarily problematic." There's no reason to think that wild animals have evolved to live in the best possible situation, just as humans aren't living their best lives when suffering from natural diseases like malaria, losing their homes and families from natural disasters, or facing a natural food shortage due to inclement weather or other circumstances beyond their control. We afford individual animals this compassion, as seen with wildlife rescue and rehabilitation. We just fail to think of the masses of wild animals from this individual-focused perspective, instead abstracting them into nonsentient entities like species and ecosystems.

Of course, even if we accept the individual-focused, interventionist philosophical view, we face a separate empirical question of how to make a positive difference in their lives. Trying to help anyone stuck in a complex situation—human or nonhuman—can backfire and cause more harm than good. This is why the field of wild animal welfare has started with research into finding effective strategies for helping these animals, and with advocacy to grow the field and get more people working on this project. Just as we would when breaching a new human issue, like a spreading pandemic that's threatening the welfare of millions of people, we should proceed with cautious urgency, collecting and assessing evidence so we can take action when we have promising interventions we can implement.

Uncertainty doesn't stop us from intervening to help humans in desperate need, and it shouldn't stop us from helping wild animals either.

Some suggest we do have lots of evidence on the likely effect of interventions to prevent wild animal suffering. Humans have devastated wildlife habitats and populations for most of our history, so this means trying to help wild animals would probably lead to negative outcomes. However, the difference in motivation and methods make this comparison very tenuous. We have almost exclusively seen the natural world as a tool for human benefit, so how could we have expected to benefit wild animals? Instead, we must judge intervention on its own merits, and if all of the interventions we come up with through careful research—including analysis of when previous interventions have used similar methods—look like they will do more harm than good, then by all means, let's not intervene.

If your intuitions are still shouting that this is messing with Mother Nature and we should steer clear, consider the perspective of the individual wild animal, just as we did with individual farmed animals at the beginning of this book. If you were an elephant or a mouse suffering in a drought—constantly living with the pain of thirst and hopelessly searching through the grasslands—wouldn't you be happy if a human filled the local watering hole? If you were suffering from an infectious disease that caused a slow, painful death, wouldn't you be happy for medication? For that matter, if someone had the opportunity to provide that assistance to you and refused to do so because they didn't want to interfere with the so-called "circle of life," would you accept that, tell them to carry on, and die your slow death, or would you desperately plead for the water or medication they hold in their hands?

To be clear, I'm not saying we should go out tomorrow and deliver all the water and vaccines we can to wild animals. I'm saying that as our moral circle expands, we should recognize the interests of these animals and conduct a rigorous evaluation of the upsides and downsides of intervention. We should study all the effects of water provision and vaccination, such as whether it will increase population size and how that will affect the welfare of other animals, as well as other methods that can be used in addition to these or on their own. We should be much more thoughtful in the actions we are currently taking that impact wild animals, evaluating how things like climate change and agriculture affect not just the global ecosystem, but also individual wild animal welfare. Perhaps these projects will have to wait until the end of animal farming and other more apparent causes of suffering, but in the meantime, we can conduct research into and raise awareness of wild animal welfare to pave the way for future advocates.[8]

Bugs

We're already in challenging territory, but I want to stretch your moral intuitions even further by considering the case of the tiny animals who live all around us. From spiders to earthworms, we give little thought to these creatures other than to swat at the occasional bothersome mosquito or take a spider out of the house.

However, bugs—a term I'm using to refer to all the small invertebrates like insects, spiders, and earthworms—are frequent subjects of academic inquiry, including from neuroscientists and biologists who have studied their nervous systems and behavior. I won't dive all the way back into a discussion of sentience, but it's safe to say that bugs show many of the behaviors we associate with sentience in our own lives, such as fleeing from danger and moving toward food. The best explanation for these actions is that they are driven by emotions like those you experience when you perform the same actions, such as fear in the case of fleeing danger, and excitement in the case of approaching a tasty meal. Many bugs even show reinforcement learning, the ability to seek out or avoid an outcome based on previous experiences.[9]

On the anatomical level, one study found that fruit fly larvae roll away when a hot object touches them, and that this behavior is mediated by neurons similar to pain neurons in vertebrates.[10] Another found that honeybees become more pessimistic when they are agitated, just like humans do.[11] Again, sentience is a complex topic that we can't do justice to here, but suffice it to say that the evidence suggests many bugs have at least some small degree of sentience, meaning we should afford them at least some moral consideration.

This could be a very tiny amount. Say you value a bug at one one-thousandth the moral worth of a farmed animal, such as a chicken. Maybe that's still too high. Let's suppose just one one-millionth. Even if that's your valuation, consider just how many bugs there are in the world—estimates suggest somewhere around 10^{18} to 10^{19} (one to ten quintillion). Using the one one-millionth estimate, that puts the total moral worth of bugs at approximately five to fifty times greater than the total moral worth of farmed animals.[12] While this is flimsy and simplified reasoning, it at least suggests that, on the whole, bugs deserve thoughtful consideration.

Most of these bugs are wild animals, so we should have the same sort of concerns about the tractability of helping them as we have about helping wild elephants and mice. However, bugs could still face significant human-caused suffering, if, for instance, insect-based foods increase in popularity. When I go to conferences and events on the future of food, insect protein is a frequent discussion topic, and there are certainly plenty of foodies who see a big role for it in the future of food. While I appreciate that this food system could reduce some of the harms of conventional animal farming, such as greenhouse gas pollution, the number of insects that would need to suffer and die for a pound of protein is many times the number of cows and pigs, and even fish and chickens. This should make us cautious about a switch to insect consumption because of the greater number of animals involved. One mitigating factor here is that insects might be happier with the confinement aspect of factory farming than mammals, birds, and fish are, but the other aspects of factory farming, like disease and slaughter, are still plenty worrisome.

Fortunately, eating insects or other bugs is not very popular at this time and there's a significant yuck factor associated with insect consumption as compared with plant-based and cultured food. Also, most experts on food ethics agree that plant-based and cultured meat are clearly much better options, even if some think eating insects would still be preferable to the current situation.

Artificial sentience

I have one final exercise for your moral reasoning: sentient beings who are nonnatural, meaning that instead of evolving the way humans and other animals have, they are adapted or created by humans or other intelligent beings. Philosophers have considered the possibility that humans will one day create sentient lifeforms. While we won't fully explore the logistics of artificial sentience, let me provide you with one thought experiment to make the case for this possible technological advancement:

Imagine that sometime in the not-so-distant future you develop a brain disease like Alzheimer's, but scientists have developed a neurosurgical technique to prevent further damage. They plan to replace potentially affected neurons in your brain with tiny computer chips that are functionally identical to healthy neurons. At first, the treatment switches out just a few neurons, but your condition deteriorates so they continue replacing your neurons one by one. With each replacement, you remain unchanged in everything you feel, think, and do. Eventually, they replace the very last neuron in your brain. The technology is highly advanced and the surgeons are highly skilled, so you are still exactly the same in all of your behavior and mental processes as you were before any neurons were replaced, except that those processes are now being run on a silicon substrate instead of a biological carbon-based substrate. In other words, you are you to every outside observer, as well as to yourself. In this situation, wouldn't it be appropriate to say you are most likely still sentient and that you shouldn't be treated any differently than you were before the surgery?

Situations like this seem theoretically possible, and though this is well beyond our current technological capacities, this possibility of mind-switch technology means that the number of artificial beings that exist in the future could be quite large. Humans could be motivated to do this for a variety of reasons. Computer equipment is very efficient, transmitting signals far faster than the human brain. It's less susceptible to deterioration over time from disease or aging. You could even make copies of your brain to team up and do all the exciting things you've always wanted to do but never had the time.[13]

Whole-brain emulation is just one of the ways so-called digital sentience could come to exist. There are many others, for instance employers could find that digital minds, driven by reinforcement learning, are better and cheaper employees than humans. Some of these minds could be far less sophisticated than human minds and made to specialize in specific tasks. They could form the labor force of a company that wants to run millions of scientific trials. They could be robotic personal servants, built to feel human feelings in order to understand us better and ensure that all our wants and needs are satisfied. The stars are the limit as human technological capacity grows. If this seems far-fetched, remember that the smartphone in your pocket has more computing power than all of NASA did in 1969 when the first humans landed on the moon.[14]

If we do create sentient machines, which many scientists see as a legitimate possibility, we could see these beings being subjected to an immense amount of suffering. Less powerful digital minds could be treated as lower classes, similar to how humans today treat animals as tools and property. In fact, if digital sentience emerges, we could see brand-new social movements emerge with these machines to fight against their oppression, just as we've seen for biological victims.

Implication: Focus on animal protection

How exactly the moral circle should expand is hard to say, but it doesn't seem like domesticated animals should be the final frontier. We should consider what effects we might have today on the moral circle of our descendants. It's not a guarantee that all sentient beings will have their moral interests fully accounted for—the vast majority alive today don't.

This is why the way the movement promotes and contextualizes animal-free food is so important. If it ties the rise of animal-free foods in society with concern for farmed animals, it sets major precedent for caring about the large-scale suffering of politically powerless sentient beings, the ones most at risk of being treated poorly in the future.

If the end of animal farming is instead tied entirely to other motivations like the environment or human health, it sets a precedent for caring more about those issues, which already have far more people working on them. These values could even conflict with the welfare of sentient beings. A stronger prioritization of human health, for example, could lead people to stop eating red meat and instead eat chickens and fish, which would likely cause more animal suffering. In the long run, a stronger focus on human health could potentially increase cruel animal experimentation if that is expected to be beneficial to human welfare.

As discussed in the section on wild animals, the conservation of ecosystems, biodiversity, and other nonsentient entities is only a proxy for the welfare of sentient beings. Intrinsic concern for Earth's current natural environment can actually harm wild animals, such as when humans release biological weapons like poison or diseases on members of invasive species in order to preserve the integrity of local ecosystems—like a plan by the Australian government to release an Israel-based herpes virus to decimate the country's carp population.[15] Preserving the integrity of an ecosystem could be good for the animals therein, of course, but that impact on animal welfare is already accounted for if your focus is directly on their welfare, and not simply on the preservation of ecosystems.

Advocates should ensure that the social change we create promotes our true values as much as possible, rather than related values such as preservationism that overlap in some, but not all, contexts.[16] One specific avenue for increasing the welfare of sentient beings is promoting the idea of legal personhood for animals, such as the work of the Nonhuman Rights Project, discussed in chapter 1. Just as being treated as a legal person instead of legal property has been a huge stride for oppressed humans, it could greatly benefit nonhumans.

This reasoning also suggests that we should take care not to privilege those on the borders of the moral circle over those farther outside it. I have been careful in this book to avoid moral comparisons of the animals we eat that raise these animals' status by diminishing the moral status of other populations, such as bugs or robots. An inclusive approach to potential harms suffered by all sentient beings is crucial to the progress of the moral circle.

Other ways to affect the far future

Ensuring the expansion of the moral circle—or at least preventing it from slowing down or stagnating, which may be a more important risk—isn't the only way we can help sentient beings in the far future. The moral quality of the universe depends not only on the welfare of sentient beings, but also on how many of them exist. For example, a survey of artificial intelligence researchers suggested a 5 percent chance of human extinction as a result of AI, meaning no sentient beings would exist.[17] As discussed earlier, devoting more resources to avoiding such an outcome could be highly cost-effective if one believes these sentient beings would be living good, happy lives if only we prevented extinction.

Building a good far future not only depends on humanity's values and existence, but also on the ability of existing humans to implement their values. In other words, can they successfully shape the universe into the type of place in which they want to exist? One crucial capability that enables people to implement their values is rationality, the ability to form accurate beliefs about the world and to make decisions that best achieve one's goals. If we believe humanity, on average, has fundamentally good goals, then increasing its ability to achieve those goals could be highly cost-effective.

A full analysis of these strategies is beyond the scope of this book, but one key reason to favor moral circle expansion over strategies like these is that merely creating greater power for humans to achieve their goals has the potential to cause not only great progress but great harm. As a reader of this book, you're familiar with at least some of the terrible suffering humanity has caused, and while much value has come out of technologies like medicine, nuclear energy, and more efficient manufacturing, these same technologies are also responsible for biological and nuclear weapons, sweatshops, and factory farms.

The end of animal farming

The end of animal farming is one key step in the broader project of moral circle expansion, and we have spent most of this book exploring the most effective strategies being used to reach that goal. But what exactly will the transition to an animal-free food system look like?

We can't know for sure, but my best guess is that there will be two initial areas of expansion for animal-free foods.[18] First, some animal-free foods, especially plant-based, relatively easy-to-make products like eggless mayonnaise, will replace animal-based foods at institutions like schools and catering companies. These purchasers are less constrained than grocery stores by public demand, so they can be won over with relative ease by the low costs of products made from plant ingredients relative to those made from animal ingredients, and ethical mandates like company-wide sustainability goals. This institutional uptake will be driven by leaders like Houman Shahi, a young entrepreneur who founded Earthrise in 2016 to help connect plant-based companies like Hungry Planet with k–12 schools. Both of these parties are limited in their ability to make these connections themselves. Plant-based companies are new and focused on product; k–12 schools are cash-strapped and embedded in the status quo. Currently, Earthrise's main focus is on chicken nuggets, a particularly harmful animal product that's particularly easy to replace with plant-based protein. While the organization has yet to make a serious dent due to its short existence and small size (just four staff), it is building a business and policy framework for a much wider replacement across US institutions.[19]

Second, early adopters who consume more expensive animal products, such as at high-end restaurants, will switch those products out for animal-free foods with similar price, such as the first cell-cultured ground beef to hit the market. These wealthier consumers tend to make their purchasing decisions based more on health and ethics than the average purchaser. While this change will do little to directly curtail the factory farming industry, it will allow animal-free food producers to generate revenue and expand their manufacturing facilities, so they can grow to economies of scale and drive prices down. Early adoption by celebrities, entrepreneurs, and business leaders will popularize these products and boost their desirability by endowing them with high status like how meat has been a symbol of wealth and status in human history due to its limited availability.

After substantial expansion has occurred in these two areas, the animal agriculture industry will lose significant profits and have fewer resources for lobbying and marketing, though they will spend a greater portion of their resources on defending themselves. I think this will lead to a net decrease in resources, which will allow, for example, a shift toward recommendations of animal-free foods in the US Dietary Guidelines, which will facilitate mass adoption, such as by promoting these foods in public school cafeterias. By this point, the average person in the West will see animal-free foods as normal options more than as weird, imperfect alternatives for people with special dietary restrictions. We will also see more blind taste tests showing that products are indistinguishable, since without these, there would always be critics claiming otherwise. This will lead to a rapid expansion of nonprofit organizations focused on animal-free foods, reaching a scale approximately the size of the modern clean energy movement. This is especially important as the animal agriculture industry will be lashing out to maintain its hold on the marketplace. Tech giants and other Fortune 500 companies will commit to serving fewer animal-based foods like how they've committed to clean energy.

A reduction of cognitive dissonance will occur as many people stop consuming animal-based foods, allowing those vegans or near-vegans to see and condemn animal farming—even in so-called humane forms—as a moral catastrophe. The four N's of meat consumption will dissipate as eating animals becomes less normal and the evidence strengthens that it's not necessary for, or even beneficial to, human health. Activists will push for policy change that supports animal-free foods by utilizing the connections to politicians, food companies, and other institutions that they built through animal welfare reform campaigns. They might push for the taxation of animal-based foods; for negative labeling like we see on cigarette packages; for more bans on specific harmful animal farming practices; and for the reduction of animal farming subsidies and the creation of plant-based food subsidies.

I would guess that this intermediate stage of a mixed food system—where a sizable portion of meat, dairy, and egg products in high-income countries will be made from plants and cell cultures, where animal-free foods are seen as normal options along with animal-based foods, the way plant-based milks in some urban coffee shops are seen today—will arrive in ten to thirty years.[20] The cultured products, especially acellular foods like dairy and eggs, will be produced at large scale by a few big food companies, though small-scale brewery-like production will be popular for some consumers. Some of the most popular products will be blended foods that mix animal-based components with animal-free components, similar to what we already see with soy used as an extender in beef, but this blending will be seen as a health and ethical benefit, rather than as a cost savings. This period will also include a more vocal opposition to these new animal-free foods, mostly by people who see them as processed, unnatural, or untraditional. That opposition won't gain much traction outside states and regions with heavy economic focuses on animal farming, thanks to growing public support for animal-free foods generated by concerns about new avian and pig flu epidemics, environmental crises like droughts, and of course a growing concern for farmed animals due to the expanding moral circle. Even in bastions of animal farming, the harm it does to local health, the environment, and the economy will lead to increasing ill will.

From that point, I'd guess it'll take another ten to twenty years—around 2038 to 2068—for animal-free foods to make up the majority of meat, dairy, and eggs in high-income countries. We will also see mixed food systems emerge at this point in urban areas in many lower-income countries. I don't think we'll see animal farming end in regions that are as poor as the poorest countries are today, where access to food technology is limited and farmed animals can be an efficient source of food and wealth savings. Animal farming will end in those regions once their incomes rise. A big uncertainty I have about this stage is whether we will see calls for the abolition of animal farming. Will the mainstream movement seek to outlaw animal farming entirely, or merely to reform the remaining industry and switch public consumption to animal-free foods? My tentative guess is that there will not be many countries where full abolition becomes the focus of the farmed animal movement at this stage, but it might happen, at least in a few small, progressive European countries like Austria and Switzerland. In some countries, we will also see an explicit push for the end of factory farming. This will be hard to implement, and we could see large public backlash if the animal agriculture industry refuses to end its industrial practices, similar to what happened when advocates tried and failed to reform British slavery in the early 1800s, leading to a successful push for the full emancipation of British slaves.[21]

Animal farms that continue to exist in high-income countries in the latter half of the twenty-first century will mostly make specialty products with greatly reduced animal suffering. By this point, reductions in animal farming will start to have noticeable effects on human health and local environments. We'll hear touching stories about factory farms being abandoned and torn down in rural areas. The waste covering the floors of those unsanitary sheds stops being kicked into the air, hoses connected to waste runoff lakes stop spraying urine and feces into the wind, and local public health improves dramatically. People working with factory farms could also find new, better jobs in the post-animal food economy, though there will be tension here as there is with any new industrial technology like computers replacing typewriters or automobiles replacing horse-drawn carriages. The animals themselves will by and large cease to be born into this cruel industry. There won't be a decision of "What to do with all these cows?" because it will just be a matter of fewer and fewer cows being born through artificial insemination. When there are freed cows, such as when an animal farmer has a change of heart and switches to plant or cell farming, there will be plenty of compassionate sanctuaries and rescuers. We will also have a surplus of land, including some grazing land that's ill-suited for crop production. In some cases, this land could be used for especially durable crops that work in the specific climate and soil conditions, or it could be reinvigorated to make it suitable for non-durable crops. But given the surplus of food once we end animal farming (recall how much plant material is required to produce meat), we could use this land for wind power, inedible plants like switchgrass that can be used as biofuels, cellular agriculture facilities, relocating humans who currently live on arable land, relocating humans who need space as Earth becomes increasingly crowded, or even just allowing nature to take over again.

If I had to speculate, I would say that by 2100 all forms of animal farming will seem outdated and barbaric. The moral momentum of ending 99 percent of animal farming will lead our descendants to oppose even the final 1 percent, just as with other historical atrocities where, eventually, the institution was seen as inherently evil despite initial opposition being to its most egregious implementations. The cows, chickens, or other animals from formerly farmed species who remain will live on sanctuaries, where humans care for them as an exercise in compassion and a reminder of our historical transition away from animal cruelty. Though even that might not happen because of how these animals are bred to suffer, even in the best sanctuary conditions.

People eat animal products in spite of how they are produced, not because of it. As humanity gains unprecedented technological power such as a deep understanding of cell and tissue biology, we will be able to create meat, dairy, eggs, leather, and other products without the metabolic waste of biological processes like movement and brainpower. And if space colonization awaits humanity, why would we bring a system as unsustainable as animal farming with us? The system's fundamental inefficiency will end animal farming one day, regardless of our concern for animals, the environment, or human health.

■ ■ ■ ■ ■

Despite my cautious optimism, I have significant uncertainty about all these long-term predictions. I'm also deeply uncertain about the continuing expansion of the moral circle from farmed animals to new populations like wild animals and digital beings, though it remains in the back of my mind as I focus on the moral progress of our generation.

This is the eternal struggle of the advocate: we march forward with our day-to-day work, whether that's through science, activism, business, consumer choices, or any other role. There's a comfortable certainty in the achievements we see every day, but sometimes we look up to see distant goals like the end of factory farming and the end of animal farming altogether. These can seem so far away, sometimes hopelessly far away. Then every once in a while, some of us even look up to the stars and think about the future of human civilization, wondering what will happen if humanity takes control of the cosmos. The stakes can feel overwhelming, burdening advocates with a deep, inescapable feeling of "weltschmerz"—a useful word for advocates to describe their burnout, it means "mental depression or apathy caused by comparison of the actual state of the world with an ideal state."[22]

But that's not the only way to look at the situation. If you recognize the moral catastrophe of animal farming, you are a moral pioneer. As you walk with other advocates, you stand in the footsteps of people from every generation, around the world, who have fought for other Copernican leaps in the expansion of humanity's moral circle such as the inclusion of women, people of color, and people who live in distant locations.

These changemakers are our mentors. Stories of their success in our history books guide us and allow us to direct our sights confidently ahead instead of fixing our gaze on the current struggles beneath our feet. What those who came before us have achieved for those of us here now should give us resolve that the world is never how it always will be; it is always ours to change. In turn, we leave our own footsteps for future moral pioneers to follow.

We've talked about a lot of advocates in this book. There are investigators who have borne witness to the cruelty of animal farming and inspired many in our generation to commit ourselves to this issue. There are nonprofit and business leaders who wear suits and sit down with CEOs to bridge the gap between big food companies and the frontier of our moral circle. There are scientists and culinary professionals who are designing and constructing the food of the future. There are rescuers who take animals out of farms and tell their stories. There are leafleters and protesters waging months-long campaigns to effect policy change. And there are advocates who work full-time on other important issues, like human health, human rights, and climate protection, building bridges between those movements and the movement to end animal farming.

This is your community. This is the generation that will revolutionize not just the way we eat, but the way we care for the many billions of sentient beings who, until now, have suffered without relief outside of humanity's moral circle. When this tragedy weighs on you, these are all people you can turn to. The advocates profiled in this book, and those working at organizations and companies dedicated to disrupting animal agriculture, are accessible via email and phone. It would make their day to hear your story and help you get involved. If you want to help with corporate campaigns for animal-free food options and better living conditions for farmed animals, consider signing up for the Hen Heroes program of Mercy For Animals or the Fast Action Network of the Humane League, or just follow these organizations on social media. If you're an entrepreneur or biologist, discuss startup opportunities with the Good Food Institute—the organization maintains a list of business ideas that could receive funding and mentoring from its network.[23] If you're a writer or social scientist, contact me at the Sentience Institute to discuss research projects on topics like the best terminology and branding for cell-cultured meat. Maintain a mind-set of effective altruism and read the results of that research to find new, exciting ways of having a bigger impact than you ever thought possible.

Above all, make connections with other advocates. Social movements live or die based on the quality of relationships among advocates. It's easy to make friends in a dedicated community that's built on compassion like the farmed animal movement. It's a big ecosystem, and it might take meeting a few different people to find your niche, but in a burgeoning social movement like this one, there's room for everyone.

ACKNOWLEDGMENTS

I am first and foremost thankful to the effective altruists and animal advocates who work to do as much good as possible. They inspire me every day.

Nobody played a bigger role in the writing of this book than my partner, in life and in research, Kelly Witwicki. She was the first person to read almost everything I wrote. She was the editor most familiar with the research, meaning she could point out my mistakes more easily than anyone else. And she was the greatest source of insight due to her brilliance and honesty. In fact, it seems there's no better time and place than right here and right now to ask this question: Kelly, will you marry me?

My deepest thanks to my family Nancy, Jodie, Delphin, and Rigel for giving me a childhood far outside the bubble in which I currently live, and for teaching me that there were no limits to my imagination and that doing good is always more important than getting first place.

I'm immensely grateful to my agent, Stacey Glick, and my editor, Will Myers. I have some pretty crazy ideas, but they both noticed insights buried in the rough and helped me refine them for a broad audience. I'm especially grateful that they found me a literary home in Beacon Press, an incredible, mission-driven institution that combines the resources of the big press with the dedication and nuance of the small.

For convincing me I could write a book that somebody might actually read, thanks to Brian Kateman. For answering my endless questions in writing this book, thanks to Lewis Bollard, Paul Shapiro, Pei Su, and Kenny Torrella. For their early and extensive editing of the manuscript, thanks to Jacob Funnell, Jay Quigley, Cameron Meyer Shorb, Kenny Torrella, and Marianne van der Werf. For rigorous feedback later in the process, thanks to Ben Davidow, Dave Doody, Andreas Haefner, Mikko Järvenpää, Alan Levinowitz, Angelina Li, Danny Lipsitz, David Kradolfer, Lila Rieber, Houman Shahi, Mimi Tran, and Ben Wurgaft.

NOTES

Introduction

1. I use the term "animal farming" to refer to the practice of raising and killing animals for food, including wild-caught fish. "Animal agriculture" refers to the industry and companies that practice and support animal farming, including food retailers and, to a lesser extent, companies that work with animal farms, such as pharmaceutical companies that provide antibiotics for livestock. "Factory farming" refers specifically to large-scale, industrial animal farming. I use the term "animals" in the colloquial sense to refer to sentient nonhuman animals (probably excluding sponges). Biologically, humans are animals. Invertebrates are unfortunately often excluded from analyses in this book, but see chapter 9 for a section on bugs.
2. Tom Chatfield, "What Our Descendants Will Deplore About Us," BBC.com, June 27, 2014, http://www.bbc.com/future/story/20140627-how-our-descendants-will-hate-us; Ezra Klein, "What Will Our Grandchildren Hate About Us?" *Washington Post*, September 29, 2010, http://voices.washingtonpost.com/ezra-klein/2010/09/what_will_our_grandchildren_ha.html; Matt Ridley, "One Day We Will See That Meat Is Murder," *Times & Sunday Times* (London), April 24, 2017, https://www.thetimes.co.uk/article/one-day-we-will-see-that-meat-is-murder-dlr597b2c; Richard Branson, "Investing in a Cleaner Way to Feed a Hungry World," Virgin.com, August 24, 2017, https://www.virgin.com/richard-branson/investing-cleaner-way-feed-hungry-world; Bill Nye, "I Am Bill Nye and I'm Here to Dare I Say It . . . Save the World. Ask Me Anything!," *Reddit*, April 19, 2017, https://www.reddit.com/r/IAmA/comments/66ajul/i_am_bill_nye_and_im_here_to_dare_i_say_it_save/dggyrvv/; Anuradha Varma, "Vegan for Life!," *Times of India*, August 28, 2010, https://timesofindia.indiatimes.com/life-style/health-fitness/diet/Vegan-for-life/articleshow/5513024.cms.
3. Witwicki, "Sentience Institute Global Farmed & Factory Farmed Animals Estimates." Insects and other invertebrates are excluded from this estimate due to a lack of data.
4. Ibid. Approximately 103 billion to 343 billion.
5. Ibid.; Reese, "Sentience Institute US Factory Farming Estimates."
6. See discussion and citations in chapter 1, "The Expanding Moral Circle."

7. Yuval Noah Harari, "Industrial Farming Is One of the Worst Crimes in History," *Guardian*, September 25, 2015, https://www.theguardian.com/books/2015/sep/25/industrial-farming-one-worst-crimes-history-ethical-question.
8. "Diet Reform and Vegetarianism," Janice Bluestein Longone Culinary Archive at the University of Michigan, accessed November 9, 2017, https://www.lib.umich.edu/janice-bluestein-longone-culinary-archive/diet-reform-and-vegetarianism; Rebecca Fowler, "An Enema of the People," *Washington Post*, October 23, 1994, https://www.washingtonpost.com/archive/lifestyle/style/1994/10/23/an-enema-of-the-people/6d510d59–4a93–403c-a60f-2b3605deac57/; Snow, *Mechanical Vibration*, 82–93.
9. Ariel Schwartz, "This Startup Is Making Real Meatballs in a Lab Without Killing a Single Animal," *Business Insider*, November 15, 2016, http://www.businessinsider.com/memphis-meats-lab-grown-meatballs-2016–11.
10. Kenny Herzog, "Why UFC's Toughest Fighters Are Going Vegan," *Men's Journal*, March 21, 2016, http://www.mensjournal.com/sports/articles/nate-diaz-and-other-vegan-ufc-fighters-w199323; Susie East, " 'Vegan Badass' Muscle Man Pumps Iron, Smashes Stereotypes," CNN.com, July 6, 2016, http://www.cnn.com/2016/07/06/health/vegan-strongman-patrik-baboumian-germany-diet/index.html; "Venus and Serena Williams," Veganuary, https://veganuary.com/people/venus-serena-williams/; Rachel Wenzlaff, "NFL Veganism? David Carter, Griff Whalen Have Broken the Mold," NFL.com, September 28, 2016, http://www.nfl.com/news/story/0ap3000000711369/article/nfl-veganism-david-carter-griff-whalen-have-broken-the-mold.
11. Alok Jha, "Google's Sergey Brin Bankrolled World's First Synthetic Beef Hamburger," *Guardian*, August 5, 2013, https://www.theguardian.com/science/2013/aug/05/google-sergey-brin-synthetic-beef-hamburger.
12. Jillian D'Onfro, "The CEO of a Startup That Makes Fake Meat Explains Why He Didn't Sell to Google," *Business Insider*, June 1, 2016, www.businessinsider.com/pat-brown-on-why-impossible-foods-didnt-sell-to-google-2016–6.
13. Katie Fehrenbacher, "The 6 Most Important Tech Trends, According to Eric Schmidt," *Fortune*, May 2, 2016, http://fortunc.com/2016/05/02/eric-schmidts-6-tech-trends/. I will use the term "plant-based" to refer to foods made entirely without the use of animals, while the term "animal free" includes both plant-based and cultured foods.
14. Numerous visits to Google's headquarters and conversations with staff, 2016.
15. Some farmed animals also eat other animals, even sometimes of their own species. See Michael Pollan, "Power Steer," *New York Times*, March 31, 2002, http://www.nytimes.com/2002/03/31/magazine/power-steer.html.
16. "Feed:Meat Ratios," A Well-Fed World, October 26, 2015, http://awfw.org/feed-ratios/.
17. For more information on why effective altruists should prioritize helping

animals, see Jacy Reese, "Why Animals Matter for Effective Altruism," Effective Altruism Forum, August 22, 2016, http://effective-altruism.com/ea/100/why_animals_matter_for_effective_altruism/.

1: The Expanding Moral Circle

1. Opposition to animal farming takes many forms depending on one's values, but in my experience a deep investigation of the harms reveals the animal welfare argument to be the most persuasive, and this is the case for the entrepreneurs and activists in this book working to promote animal-free food.
2. Spencer, *The Heretic's Feast*, 201; Cottingham, "'A Brute to the Brutes?'"
3. See the first recorded discussion of the *scala naturae* in Aristotle, *History of Animals*.
4. Dio, paragraph 25, in Book LXVI, *Roman History*.
5. Pinker, "Cruel and Unusual Punishments," in *The Better Angels of Our Nature*.
6. Mandy Zibart, "Who Are Change.org's 100 Million Users?," Change.org, June 23, 2015, https://www.change.org/l/us/who-are-the-100-million.
7. Rebecca Rifkin, "In U.S., More Say Animals Should Have Same Rights as People," Gallup, May 18, 2015, http://www.gallup.com/poll/183275/say-animals-rights-people.aspx.
8. Chris Berry, "All 50 States Now Have Felony Animal Cruelty Provisions!" Animal Legal Defense Fund, March 14, 2014, http://aldf.org/blog/50-states-now-have-felony-animal-cruelty-provisions/.
9. Humane Society, "Welfare Issues with Gestation Crates for Pregnant Sows."
10. Benson, "Advancing Aquaculture," 15–16.
11. Humane Society, "Welfare Issues with Selective Breeding of Egg-Laying Hens for Productivity."
12. "Nicolaus Copernicus," *Stanford Encyclopedia of Philosophy*, November 30, 2004, http://plato.stanford.edu/entries/copernicus/.
13. Darwin, *The Expression of the Emotions in Man and Animals*, 36, 127.
14. Jörgensen, "Empathy, Altruism and the African Elephant."
15. Sato et al., "Rats Demonstrate Helping Behavior Toward a Soaked Conspecific."
16. Marino and Colvin, "Thinking Pigs."
17. Berndorff, "Klug, aber schweinisch."
18. This model has been criticized as lacking an empirical basis, but its persistence in the public zeitgeist makes it a useful example for my purposes here.
19. "Echo: An Elephant to Remember," *Nature*, October 14, 2008, http://www.pbs.org/wnet/nature/echo-an-elephant-to-remember-elephant-emotions/4489/.
20. Sacks, section 7.8, in *An Anthropologist on Mars*.
21. Matt McCall, "Watch: Octopus Carries Coconut—But Is It Using a Tool?" *National Geographic*, June 5, 2015, http://news.nationalgeographic.com/2015/06/150605-octopus-tools-animals-ocean-science/.

22. Darwin, *The Formation of Vegetable Mould*, 23–25.
23. Obee and Ellis, *Guardians of the Whales*.
24. Safina, *Beyond Words*, 27.
25. De Waal, *Are We Smart Enough to Know How Smart Animals Are?*, 9.
26. Safina, *Beyond Words*, 27.
27. Masson, *When Elephants Weep*, 11.
28. Herzog, *Some We Love*, 214.
29. Low, "Cambridge Declaration on Consciousness."
30. "SeaWorld Responds to Questions About Captive Orcas, 'Blackfish' Film," CNN.com, October 28, 2013, http://www.cnn.com/2013/10/21/us/seaworld-blackfish-qa/.
31. Tim Zimmerman, "The Killer in the Pool," *Outside*, July 30, 2010, https://www.outsideonline.com/1924946/killer-pool.
32. SeaWorld's current statement on *Blackfish* is at https://seaworldcares.com/the-facts/truth-about-blackfish/, accessed November 9, 2017. See also Jenny Kutner, "The 5 Dumbest Things SeaWorld Has Done in Response to 'Blackfish' (So Far)," *Dodo*, April 23, 2014, https://www.thedodo.com/the-5-dumbest-things-seaworld—521507954.html.
33. Maya Rhodan, "Seaworld's Profits Drop 84% After Blackfish Documentary," *Time*, August 6, 2015, http://time.com/3987998/seaworlds-profits-drop-84-after-blackfish-documentary/.
34. "Q&A about the Nonhuman Rights Project," Nonhuman Rights Project, https://web.archive.org/web/20160811114634/http://www.nonhumanrightsproject.org/qa-about-the-nonhuman-rights-project, accessed November 9, 2017.
35. Michael Mountain, "Update on Hercules and Leo," Nonhuman Rights Project, *Nonhuman Rights Blog*, July 31, 2015, http://www.nonhumanrightsproject.org/2015/07/31/update-on-hercules-and-leo/.
36. Serpell, "How Social Trends Influence Pet Ownership." It's unclear what the role of homeless and stray animals is in this statistic.
37. Kellert, "American Attitudes Toward and Knowledge of Animals."
38. Rothgerber and Mican, "Childhood Pet Ownership, Attachment to Pets, and Subsequent Meat Avoidance."
39. Gabe Bullard, "The World's Newest Major Religion: No Religion," *National Geographic*, April 22, 2016, http://news.nationalgeographic.com/2016/04/160422-atheism-agnostic-secular-nones-rising-religion/.
40. Jane Hughes, "The Best Countries in the World for Vegetarians," *Guardian*, September 23, 2013, http://www.theguardian.com/lifeandstyle/wordofmouth/2013/sep/23/best-countries-to-be-vegetarian.
41. Stephanie van den Berg, "Beatles Guru Maharishi Mahesh Yogi Dies," *Sydney Morning Herald*, February 7, 2008, https://www.smh.com.au/world/beatles-guru-maharishi-mahesh-yogi-dies-20080206-1qno.html.
42. "The Beatles (1960–70)," International Vegetarians Union, https://ivu.org/people/music/beatles.html, accessed November 9, 2017.
43. John 10:11–18 (King James Version).
44. Gen. 1:26 (King James Version).
45. Kristin M. Swenson, "The Bible and Human 'Dominion' over Animals:

Superiority or Responsibility?" *Huffington Post*, August 14, 2010, http://www.huffingtonpost.com/kristin-m-swenson-phd/the-bible-and-human-domin_b_681363.html.

46. "Does God Desire Animal Sacrifices?," *The Skeptic's Annotated Bible*, http://skepticsannotatedbible.com/contra/desire.html, accessed November 9, 2017.
47. Charlotte Meredith, "Thousands of Animals Have Been Saved in Nepal as Mass Slaughter Is Cancelled," *Vice*, July 29, 2015, https://news.vice.com/article/thousands-of-animals-have-been-saved-in-nepal-as-mass-slaughter-is-cancelled.

2: Emptying the Cages

1. Reese, "Sentience Institute US Factory Farming Estimates."
2. Arndt, *Battery Brooding*.
3. Gina-Marie Cheeseman, "California Law Banning Confinement Crates Takes Effect in 2015," *TriplePundit*, December 31, 2014, https://www.triplepundit.com/2014/12/california-law-banning-confinement-crates-takes-effect-2015/.
4. Michael Goldberg, "The Failure of Prop 2 in California," *Daily Pitchfork*, February 25, 2016, http://dailypitchfork.org/?p=1040; Wayne Pacelle, "Breaking News: First-Ever Criminal Charges Brought Against Egg Producer for Violating California Prop 2," Humane Society of the United States, February 7, 2017, http://blog.humanesociety.org/wayne/2017/02/breaking-news-first-ever-criminal-charges-brought-egg-producer-violating-california-prop-2.html; conversation with an animal welfare advocate who worked with a local government official on a police raid of a factory farm to see if the ban was in effect. The raid was necessary because the factory farmer would not allow the official to visit, but it was not publicized. The estimate of twelve million is from "About the U.S. Egg Industry," American Egg Board website, http://www.aeb.org/farmers-and-marketers/industry-overview, accessed November 9, 2017.
5. "Meat Boycott Spreads over United States," *San Francisco Call*, January 22, 1910, https://cdnc.ucr.edu/cgi-bin/cdnc?a=d&d=SFC19100122.2.2.
6. Smith, "Factory Farms."
7. Leah Garces, "Dead-End Genetics: Why the Chicken Industry Needs a New Roadmap," *Food Safety News*, January 15, 2014, http://www.foodsafetynews.com/2014/01/dead-end-genetics-why-the-chicken-industry-needs-a-new-roadmap/.
8. Williams, *Delmarva's Chicken Industry*, 11, 23.
9. Lebergott, "Labor Force and Employment, 1800–1960"; US Department of Labor, Bureau of Labor Statistics, "Employment by Major Industry Sector," https://www.bls.gov/emp/ep_table_201.htm.
10. Matthew Prescott, "Your Pig Almost Certainly Came from a Factory Farm, No Matter What Anyone Tells You," *Washington Post*, July 15, 2014, https://www.washingtonpost.com/posteverything/wp/2014/07/15/your-pig-almost-certainly-came-from-a-factory-farm-no-matter-what-anyone-tells-you/.

11. Paul Shapiro, "The Elephant-Sized Subsidy in the Race," *National Review*, February 17, 2016, http://www.nationalreview.com/article/431453/end-farm-subsidies-now.
12. Knoll, "Origins of the Regulation of Raw Milk Cheeses in the United States"; Albala, "Big Business and the Homogenization of Food," in *Food*.
13. "Guns," PollingReport.com, http://www.pollingreport.com/guns.htm, accessed November 9, 2017.
14. Josh Israel, "The Gun Industry Has Systematically Demolished Regulators and Avoided the Fate of Cigarettes," *ThinkProgress*, December 1, 2015, https://thinkprogress.org/the-gun-industry-has-systematically-demolished-regulators-and-avoided-the-fate-of-cigarettes-c3a763da0b16.
15. Jill Richardson, "ALEC Exposed: Protecting Factory Farms and Sewage Sludge?" Center for Media and Democracy's PR Watch, August 4, 2011, http://www.prwatch.org/news/2011/08/10922/alec-exposed-protecting-factory-farms-and-sewage-sludge.
16. American Egg Board, *Annual Report 2015*, 30.
17. Election 2016 results tool, "Massachusetts Ballot Question 3: Improve Farm Animal Confines," *Boston Globe*, https://www.bostonglobe.com/elections/2016/MA/Question/3%20-%20Improve%20Farm%20Animal%20Confines.
18. Email from Jenny Woods, senior communications administrator at PETA, January 10, 2017.
19. "The Grief Behind Foie Gras Duck and Goose Liver Pate," All-Creatures.org, http://www.all-creatures.org/articles/foiegras-peta.html, accessed November 9, 2017.
20. Case no. 1:14-CV-104 civil rights complaint, http://aldf.org/wp-content/uploads/2015/08/3–17–14-ALDF-complaint-ag-gag.pdf, accessed November 9, 2017.
21. "PETA's McCruelty Campaign Timeline," People for the Ethical Treatment of Animals, http://www.mccruelty.com/why.aspx.
22. To appreciate the downsides of animal costumes in activism, consider how offensive and trivializing it would be to dress up as any oppressed human group, even those who usually can't participate in protest themselves, such as the global poor or young children. While the case of animals is different in some ways, costumes still make the cause look less serious and less important.
23. Animal Charity Evaluators, "Exploratory Review: PETA." The ineffectiveness of PETA's use of sexually explicit imagery in its campaigns has even been evidenced by at least one scientific trial. See Bongiorno, Bain, and Haslam, "When Sex Doesn't Sell." There is also evidence in business marketing that "sex sells" is not a productive strategy, despite being attention-grabbing. See Wirtz, Sparks, and Zimbres, "Effect of Exposure to Sexual Appeals"; Spencer Chen, "Booth Babes Don't Work," *TechCrunch*, January 13, 2014, https://techcrunch.com/2014/01/13/booth-babes-dont-convert/.
24. Ingrid Newkirk, "Hey, What If Factory Farming Were THE ONLY THING Anyone Worked to End?," Animal Rights National Conference,

published online September 1, 2016, https://www.peta.org/blog/ingrid-newkirk-animal-rights-conference-speech/.

25. Lori Montgomery, "Md. Egg Farm Accused of Cruelty," *Washington Post*, June 6, 2001, https://www.washingtonpost.com/archive/local/2001/06/06/md-egg-farm-accused-of-cruelty/e6db2333–3b49–480a-85c0–672c7b9d08cd/.
26. Nathan Runkle, "How One Piglet Changed My Life," Forks Over Knives, July 9, 2015, https://www.forksoverknives.com/ally-to-animals-mercy-for-animals/.
27. Animal Charity Evaluators, "2015 Comprehensive Review: Mercy For Animals."
28. Animal Charity Evaluators, "Why Farmed Animals?"
29. Peter Spendelow, "Our Kinship with Animals: Wayne Pacelle Presents at VegFest," Northwest VEG, August 30, 2011, http://nwveg.org/news?entry=258.
30. "U.S. Orders Largest Ever Beef Recall," CBSNews.com, February 17, 2008, http://www.cbsnews.com/news/us-orders-largest-ever-beef-recall/.
31. "Tyson Tortures Animals," Mercy For Animals, http://www.tysontorturesanimals.com/; Helen Regan, "Foster Farms Suspends Five Employees After Graphic Video Uncovers Animal Abuse," *Time*, June 18, 2015, http://time.com/3925835/foster-farms-mercy-for-animals-abuse-chickens-investigation-graphic-video/; Stephanie Strom, "How 'Cage-Free' Hens Live, in Animal Advocates' Video," *New York Times*, October 20, 2016, https://www.nytimes.com/2016/10/21/business/video-reveals-how-cage-free-hens-live-animal-advocates-say.html.
32. SF 431 (Iowa 2011), http://coolice.legis.iowa.gov/Cool-ICE/default.asp?Category=BillInfo&Service=oldbillbook&ga=84&hbill=SF431&menu=text, accessed November 9, 2017.
33. Mark Bittman, "Who Protects the Animals?," *New York Times*, April 16, 2011, https://opinionator.blogs.nytimes.com/2011/04/26/who-protects-the-animals/.
34. Luke Runyon, "Judge Strikes Down Idaho 'Ag-Gag' Law, Raising Questions for Other States," NPR, August 4, 2015, https://www.npr.org/sections/thesalt/2015/08/04/429345939/idaho-strikes-down-ag-gag-law-raising-questions-for-other-states; Bill Chappell, "Judge Overturns Utah's 'Ag-Gag' Ban on Undercover Filming at Farms," NPR, July 8, 2017, https://www.npr.org/sections/thetwo-way/2017/07/08/536186914/judge-overturns-utahs-ag-gag-ban-on-undercover-filming-at-farms.
35. Robbins et al., "Awareness of Ag-Gag Laws Erodes Trust in Farmers."
36. Sally Jo Sorensen, "In Farmfest Panel, MN Livestock Producer Group Reps Repudiate ALEC-Inspired 'Ag Gag' Bill," *Bluestem Prairie*, August 7, 2013, http://www.bluestemprairie.com/bluestemprairie/2013/08/in-farmfest-panel-mn-livestock-producer-group-reps-repudiate-alec-inspired-ag-gag-bill.html.
37. Conversation with David Coman-Hidy, 2017.
38. Animal Charity Evaluators, "2015 Comprehensive Review: Mercy For Animals." These estimates don't account for the costs of enforcement:

monitoring companies to ensure they have made these changes in their supply chains, campaigning against any companies that don't follow through, eventually passing legislative reforms that solidify the new practices, and finally enforcing those new laws.

39. Bollard, "Initial Grants"; Blokhuis and Arkes, "Some Observations on the Development of Feather-Pecking"; Perry, "Cannibalism," in *Welfare of the Laying Hen*.
40. United Egg Producers homepage, http://www.unitedegg.org/.
41. Krissa Welshans, "The Cage-Free Egg Dilemma," *Feedstuffs*, March 7, 2016, http://fdsmagissues.feedstuffs.com/fds/PastIssues/FDS8803/fds08_8803.pdf.
42. Bollard, "Initial Grants."
43. Sentience Institute, "Momentum vs. Complacency from Welfare Reforms."
44. Witwicki, "Social Movement Lessons."
45. Mercy For Animals, "Why We Work for Policy Change"; Mullally and Lusk, "The Impact of Farm Animal Housing Restrictions."
46. Mercy For Animals, "Why We Work for Policy Change."
47. Ibid.
48. Animal Equality, iAnimal, http://ianimal360.com/.
49. "How Many Vegetarians Are There?," *Vegetarian Journal*, September/October 1997, https://www.vrg.org/journal/vj97sep/979poll.htm; "Here's Who We Are," *Vegetarian Times*, October 1992.
50. Vegetarian Research Group FAQ, http://www.vrg.org/nutshell/faq.htm#poll.
51. Frank Newport, "In U.S., 5% Consider Themselves Vegetarians," Gallup, July 26, 2012, http://news.gallup.com/poll/156215/consider-themselves-vegetarians.aspx.
52. Table 5.1, in 2014 India Census, http://www.censusindia.gov.in/vital_statistics/BASELINE%20TABLES07062016.pdf.
53. Reese, "Survey of US Attitudes Towards Animal Farming."
54. Humane Society, "Farm Animal Statistics."
55. American Humane Association, "2014 Humane Heartland Farm Animal Welfare Survey."
56. Reese, "Survey of US Attitudes Towards Animal Farming."

3: The Rise of Vegan Tech

1. Elaine Watson, "Plant Egg Entrepreneur: 'We're Not in Business Just to Sell Products to Vegans in Northern California,'" *FoodNavigator-USA*, September 12, 2013, http://www.foodnavigator-usa.com/People/Plant-egg-entrepreneur-We-re-not-in-business-just-to-sell-products-to-vegans-in-Northern-California.
2. Christine Birkner, "The Good Egg," *Marketing News*, September 2013, http://www.amainsights.com/Content/BooksContents/102/MN_Sept2013-lr.pdf.
3. By one estimate, three million to four million cars by 1880, "long before the horse population reached its peak." See Eric Morris, "From Horse Power to Horsepower," *Access*, Spring 2007, https://web.archive.org/web/20170402105438/http://www.uctc.net/access/30

/Access%2030%20-%2002%20-%20Horse%20Power.pdf; Stephen Davies, "The Great Horse-Manure Crisis of 1894," Foundation for Economic Education, September 1, 2004, https://fee.org/articles/the-great-horse-manure-crisis-of-1894/.

4. US Food & Drug Administration, "Hampton Creek Foods 8/12/15: Warning Letter," August 12, 2015, https://www.fda.gov/ICECI/EnforcementActions/WarningLetters/2015/ucm458824.htm.
5. Craig Giammona and Anna Edney, "'Just Mayo' Startup Keeps Product Name Despite Lack of Eggs," *Bloomberg*, December 17, 2015, https://www.bloomberg.com/news/articles/2015–12–17/-just-mayo-startup-will-keep-product-name-despite-lack-of-eggs.
6. Sam Thielman and Dominic Rushe, "Government-Backed Egg Lobby Tried to Crack Food Startup, Emails Show," *Guardian*, September 2, 2015, https://www.theguardian.com/us-news/2015/sep/02/usda-american-egg-board-hampton-creek-just-mayo.
7. Olivia Zaleski, Peter Waldman, and Ellen Huet, "How Hampton Creek Sold Silicon Valley on a Fake-Mayo Miracle," *Bloomberg*, September 22, 2016, https://www.bloomberg.com/features/2016-hampton-creek-just-mayo/.
8. Beth Kowitt, "Feds Said to Close Inquiries into Hampton Creek's Mayo Buybacks," *Fortune*, March 24, 2017, http://fortune.com/2017/03/24/hampton-creek-sec-doj-inquiries-closed/.
9. "Hampton Creek on Buying Up Its Own Mayo," *New York Times*, August 5, 2016, https://www.nytimes.com/video/business/dealbook/100000004571839/hampton-creek-on-buying-up-its-own-mayo.html.
10. Gabler, *Walt Disney*, 139.
11. "New NPR Podcast 'How I Built This' Begins with Spanx," *National Public Radio*, September 15, 2016, http://www.npr.org/2016/09/15/494043657/how-i-built-this-spanx.
12. Chase Purdy, "A Food Tech Darling Is Tainted for Reportedly Buying Loads of Its Own Vegan Mayo," *Quartz*, August 4, 2016, https://qz.com/751140/a-food-tech-darling-is-tainted-for-reportedly-buying-loads-of-its-own-vegan-mayo/.
13. Nick Wingfield and Katie Benner, "Hampton Creek, Maker of Just Mayo, Is Said to Be Under Inquiry," *New York Times*, August 19, 2016, https://www.nytimes.com/2016/08/20/business/hampton-creek-investigation-just-mayo.html.
14. "Big Companies Without Profits—Amazon, Twitter, Uber and Other Big Names That Don't Make Money," *Fox News*, October 30, 2015, http://www.foxbusiness.com/markets/2015/10/30/big-companies-without-profits-amazon-twitter-uber-and-other-big-names-that-dont.html.
15. Svati Kirsten Narula, "The FDA Warns Food Start-Up Hampton Creek: You Can't Call It 'Mayo' If It's Not Mayonnaise," *Quartz*, August 25, 2015, https://qz.com/487656/the-fda-warns-food-start-up-hampton-creek-you-cant-call-it-mayo-if-its-not-mayonnaise/.
16. Shurtleff and Aoyagi, "History of Tofu."
17. Ginny Messina, "Soyfoods in Asia: How Much Do People Really Eat?"

Vegan RD, March 1, 2011, http://www.theveganrd.com/2011/03/soy foods-in-asia-how-much-do-people-really-eat.html.

18. Shurtleff and Aoyagi, "History of Yuba"; Shurtleff and Aoyagi, "History of Tempeh."
19. Shurtleff and Aoyagi, "Madison College."
20. K. Annabelle Smith, "The History of the Veggie Burger," *Smithsonian*, March 19, 2014, http://www.smithsonianmag.com/arts-culture/history-veggie-burger-180950163/.
21. There's no survey that I know of to establish the most famous brand, but I think most US vegetarian food experts would agree with this assessment.
22. "I Can't Believe It's Not Turkey," *Vegetarian Times*, June 1998.
23. Toni Okamoto, "Celebrating 20 Years of Tofurky: An Interview with Founder Seth Tibbott," Vegan Outreach, November 21, 2014, http://veganoutreach.org/20-years-of-tofurky-an-interview-with-founder-seth-tibbott/.
24. Mosko, *Quadripartite Structures*.
25. Becca Bartleson, "Tofurky Grows in Popularity," Oregon Public Broadcasting, November 26, 2008, http://www.opb.org/news/article/tofurky-grows-popularity/.
26. Katie Hope, "The Vegan Trying to Make the Perfect Burger," *BBC News*, January 19, 2017, http://www.bbc.com/news/business-38664353.
27. Rowen Jacobsen, "The Biography of a Plant-Based Burger," *Pacific Standard*, September 6, 2016, https://psmag.com/the-biography-of-a-plant-based-burger-31acbecbodcc.
28. Brewer, "The Chemistry of Beef Flavor."
29. Jacobsen, "The Biography of a Plant-Based Burger."
30. Bastide et al., "Heme Iron from Meat and Risk of Colorectal Cancer."
31. Gabrielle Lemonier, "Great-Tasting Veggie Burgers Are Here, but Are They Any Healthier?," *Men's Journal*, November 9, 2016, http://www.mensjournal.com/food-drink/articles/great-veggie-burgers-are-here-but-are-they-any-healthier-w449490.
32. Lindsey Hoshaw, "Silicon Valley's Bloody Plant Burger Smells, Tastes and Sizzles Like Meat," NPR, June 21, 2016, http://www.npr.org/sections/thesalt/2016/06/21/482322571/silicon-valley-s-bloody-plant-burger-smells-tastes-and-sizzles-like-meat.
33. Tara Loader Wilkinson, "How the Humble Burger Could Save the Planet," *Billionaire*, May 11, 2017, http://www.billionaire.com/billionaires/bill-gates/2811/how-the-humble-burger-could-save-the-planet.
34. Serena Dai, "David Chang Adds Plant Based 'Impossible Burger' to Nishi Menu," *Eater*, July 26, 2016, http://ny.eater.com/2016/7/26/12277310/david-chang-impossible-burger-nishi.
35. Brian Niemietz, "Don't Mess with the Chef!," *New York Post*, June 9, 2010, http://nypost.com/2010/06/09/dont-mess-with-the-chef/.
36. Thomas Heath, "One of These Burgers Is Not Like the Others," *Washington Post*, October 26, 2016, http://www.chicagotribune.com/goo/business/ct-whole-foods-beyond-meat-veggie-burger-20161026-story.html.
37. Conversation with Todd and Jody Boyman, 2017.

38. "GFI Outcomes: Why GFI is a Superb Philanthropic Investment," Good Food Institute, discussion document prepared for Animal Charity Evaluators, August 12, 2016, https://animalcharityevaluators.org/wp-content/uploads/2016/11/GFI-outcomes-successes-doc-2016.pdf.
39. Wild et al., "The Evolution of a Plant-Based Alternative."
40. Krintirasa et al., "On the Use of the Couette Cell Technology."
41. "Soy and Your Health," Physicians Committee for Responsible Medicine, http://www.pcrm.org/health/health-topics/soy-and-your-health. Note that PCRM is a nonprofit that strongly endorses plant-based eating, but this is still the best collection of literature on the topic that I know of.
42. Kate Ross, "Soy Substitute Edges Its Way into European Meals," *New York Times*, November 16, 2011, http://www.nytimes.com/2011/11/17/business/energy-environment/soy-substitute-edges-its-way-into-european-meals.html.
43. Stephanie Strom, "Trade Group Lobbying for Plant-Based Foods Takes a Seat in Washington," *New York Times*, March 6, 2016, https://www.nytimes.com/2016/03/07/business/trade-group-lobbying-for-plant-based-foods-takes-a-seat-in-washington.html.
44. Stephanie Strom, "General Mills Adds Kite Hill to Food Start-Up Investments," *New York Times*, May 19, 2016, https://www.nytimes.com/2016/05/20/business/general-mills-adds-kite-hill-to-food-start-up-investments.html.
45. Jade Scipioni and Matthew V. Libassi, "Tyson Foods CEO: The Future of Food Might Be Meatless," *Fox News*, March 7, 2017, http://www.foxbusiness.com/features/2017/03/07/tysons-new-ceo-future-food-isnt-meat.html.
46. Ethan Brown, "Why I Am Welcoming Tyson Foods as an Investor to Beyond Meat," Beyond Meat, October 10, 2016, http://beyondmeat.com/whats-new/view/why-i-am-welcoming-tyson-foods-as-an-investor-to-beyond-meat.
47. Hellmann's product page, accessed November 9, 2017, http://www.hellmanns.com/product/detail/1110074/vegan-carefully-crafted-dressing.
48. Maria Yagoda, "Veggie Burgers That Actually 'Bleed' Sold Out in an Hour at Whole Foods," *People*, May 26, 2016, http://people.com/food/veggie-burgers-that-actually-bleed-sold-out-in-an-hour-at-whole-foods/.
49. Jacy Reese, "Even at Whole Foods' Best Farm, Turkeys Suffer," *Huffington Post*, November 25, 2016, https://www.huffingtonpost.com/jacy-reese/even-at-whole-foods-best_b_8648312.html.
50. Wegmans store locator and product webpages, https://www.wegmans.com/, accessed November 12, 2017.
51. Imogen Rose-Smith, "Private Equity Veteran Jeremy Coller Champions Farm Animal Welfare," *Institutional Investor*, January 14, 2016, http://www.institutionalinvestor.com/article/3521168/asset-management-hedge-funds-and-alternatives/private-equity-veteran-jeremy-coller-champions-farm-animal-welfare.html.
52. FAIRR website, http://www.fairr.org/, accessed November 12, 2017.
53. "Tyson Investors Call for Environmental, Social Changes," *Meat+Poultry*,

August 24, 2016, http://www.meatpoultry.com/articles/news_home/Business/2016/08/Tyson_investors_call_for_envir.aspx?ID=%7B4E28BCD7-045D-489C-8A41-48A6DDDBE99F%7D.

4: How Plant-Based Will Take Over

1. Tobey, *Plowshares*.
2. Rachel Manteuffel, "Meet the Guy Who Envisions a 'Meat Brewery' to Help Solve a Global Problem," *Washington Post*, July 28, 2016, https://www.washingtonpost.com/lifestyle/magazine/meet-the-guy-who-envisions-a-meat-brewery-to-help-solve-a-global-problem/2016/07/22/2e5716b8-3e30-11e6-a66f-aa6c1883b6b1_story.html; Bruce Friedrich, "Nerds over Cattle: How Food Technology Will Save the World," *Wired*, October 7, 2016, https://www.wired.com/2016/10/nerds-cattle-food-technology-will-save-world/; Gigen Mammoser, "Why the Meat Factories of the Future Will Look Like Breweries," *Vice*, October 5, 2016, https://munchies.vice.com/en_us/article/nzk43b/why-the-meat-factories-of-the-future-will-look-like-breweries; Jessica Roy, "Thousands Petition to Get In-N-Out to Add a Veggie Burger to the Menu," *Los Angeles Times*, September 22, 2016, http://www.latimes.com/food/dailydish/la-dd-in-n-out-vegetarian-option-petition-20160922-snap-story.html; Emily Byrd, "Will Adding a Veggie Burger to the In-N-Out Menu Destroy the Country?" *Los Angeles Times*, September 26, 2016, http://www.latimes.com/opinion/op-ed/la-oe-byrd-veggie-burder-in-n-out-20160926-snap-story.html.
3. Emily Byrd, "2016: A Tipping Point for Food," Good Food Institute, December 20, 2016, http://www.gfi.org/2016-a-tipping-point-for-food.
4. Lifespan and kg/animal are from Tomasik, "How Much Direct Suffering Is Caused by Various Animal Foods?" Kilogram to calorie conversions were based on USDA Food Composition Databases. For chickens and turkey, the whole option was used. For eggs, which required a cooking method, boiled was used. For farmed fish, salmon was used to match Tomasik's estimate; no USDA estimate for catfish was available. For beef, 85/15 and broiled was used. For pork, the whole loin, broiled option was used. For milk, 2 percent and "with added nonfat milk solids and vitamin A and vitamin D" was used. Cumulative elasticity factors are from Animal Charity Evaluators' Impact Calculator: chicken 0.63, turkey 0.28, eggs 0.82, farmed fish 0.43, beef 0.67, pork 0.76, dairy 0.31. https://docs.google.com/spreadsheets/d/1zvSHSG6QDsxJWSYNVcwTb0-5u1Ms2M6EQ-ALDM53pN0/. This calculator is no longer used by the organization.
5. Ibid.
6. Bob Goldberg and Paul Lewin, "Follow Your Heart Celebrates 40 Years! A Love Story," Follow Your Heart, April 10, 2013, http://followyourheart.com/follow-your-heart-celebrates-40-years-a-love-story/.
7. Merrill Shindler, "Love Vegetarian Food? Follow Your Heart to This Canoga Park Restaurant," *Los Angeles Daily News*, January 17, 2017, http://www.dailynews.com/arts-and-entertainment/20170117/love-vegetarian-food-follow-your-heart-to-this-canoga-park-restaurant.

8. Regan Morris, "The Vegan Boss Who Followed His Heart and Made Millions," *BBC News*, December 19, 2016, http://www.bbc.com/news/business-38248169.
9. Follow Your Heart website, accessed November 12, 2017, https://followyourheart.com/.
10. Bianca Bosker, "Mayonnaise, Disrupted," *Atlantic*, October 2, 2017, https://www.theatlantic.com/magazine/archive/2017/11/hampton-creek-josh-tetrick-mayo-mogul/540642/.
11. Morris, "The Vegan Boss Who Followed His Heart and Made Millions."
12. Conversation with Miyoko Schinner, 2017; Holly Feral, "Miyoko Schinner: The Tale of a Tenacious Entrepreneur," *Driftwood*, June 4, 2016, http://www.driftwoodmag.com/single-post/2016/06/04/Miyoko-Schinner-The-Tale-of-a-Tenacious-Entrepreneur.
13. Tara Duggan, "Vegan Cheese Startup Miyoko's Kitchen Drawing Lots of Investors," *San Francisco Chronicle*, February 16, 2017, http://www.sfgate.com/business/article/Vegan-cheese-company-draws-big-bucks-from-startup-10935279.php.
14. US Food and Drug Administration, *Code of Federal Regulations Title 21*, https://www.accessdata.fda.gov/scripts/cdrh/cfdocs/cfcfr/CFRSearch.cfm?fr=131.110.
15. Moore, "Food Labeling Regulation."
16. Pharr, *Theodosian Code*, 418.
17. Julie Butler, "Two Plead Guilty in Olive Oil Fraud Scheme, Sentenced to Two Years in Jail," *Olive Oil Times*, December 17, 2011, https://www.oliveoiltimes.com/olive-oil-basics/jail-term-for-olive-oil-fraudsters/23081.
18. Tania Branigan, "Chinese Figures Show Fivefold Rise in Babies Sick from Contaminated Milk," *Guardian*, December 2, 2008, https://www.theguardian.com/world/2008/dec/02/china; Associated Press, "China's Top Food Safety Official Resigns," *NBC News*, September 22, 2008, http://www.nbcnews.com/id/26827110; "11 Countries Stop Milk Imports from China," *New Delhi Television Limited*, September 23, 2008, https://web.archive.org/web/20080926235006/http://www.ndtv.com/convergence/ndtv/story.aspx?id=NEWEN20080066441.
19. Food and Drug Law at Keller and Heckman, "Proposed 'Dairy Pride Act' Creates Controversy," *National Law Review*, January 31, 2017, http://www.natlawreview.com/article/proposed-dairy-pride-act-creates-controversy.
20. "Got 'Milk'? Dairy Farmers Rage Against Imitators but Consumers Know What They Want," editorial, *Los Angeles Times*, January 4, 2017, http://www.latimes.com/opinion/editorials/la-ed-plant-milk-fda-20170104-story.html; Melody Hahm, "Dairy Farmers Are Losing the Battle over 'Milk,'" *Yahoo!*, December 22, 2016, http://finance.yahoo.com/news/dairy-farmers-are-losing-battle-over-the-word-milk-fda-184248591.html.
21. "US Sales of Dairy Milk Turn Sour as Non-Dairy Milk Sales Grow 9%," Mintel, April 20, 2016, http://www.mintel.com/press-centre/food-and-drink/us-sales-of-dairy-milk-turn-sour-as-non-dairy-milk-sales-grow-9-in-2015; Mary Ellen Shoup, "Dairy Industry Sees Rise of Plant-Based Milk as 'Serious Threat,'" *DairyReporter*, May 4, 2017, https://www

.dairyreporter.com/Article/2017/05/04/Dairy-industry-sees-rise-of-plant-based-milk-as-serious-threat; Scheherazade Daneshkhu, "Dairy Shows Intolerance to Plant-Based Competitors," *Financial Times*, July 14, 2017, https://www.ft.com/content/73b37e7a-67a3–11e7–8526–7b38dcaef614.

22. "Calcium and Bioavailability," Dairy Nutrition, https://www.dairynutrition.ca/nutrients-in-milk-products/calcium/calcium-and-bioavailability; Tang et al., "Calcium Absorption," accessed November 12, 2017.
23. "What's Inside a Plant-Based Burger," *The Dr. Oz Show*, January 2, 2017, http://www.doctoroz.com/episode/21-day-weight-loss-breakthrough?video_id=5257028180001.
24. Average consumption, see Pasiakos et al., "Sources and Amounts." Recommended Daily Allowance, see Institute of Medicine, table S-7, in *Dietary Reference Intakes*. Some demographics, such as young women and elderly people, are at risk for insufficient levels of protein. However, I have never seen it suggested that mistaking low-protein milk for high-protein milk would increase risk in these demographics.
25. SuperMeat website, http://supermeat.com/meat.html, accessed November 12, 2017.
26. Elaine Watson, "Game-Changing Plant-Based 'Milk' Ripple to Roll Out at Whole Foods, Target Stores Nationwide," *FoodNavigator-USA*, April 18, 2016, http://www.foodnavigator-usa.com/Manufacturers/Plant-based-milk-Ripple-to-roll-out-at-Whole-Foods-Target.
27. Ripple Foods website, http://ripplefoods.com/whatshouldmilkbe/, accessed November 12, 2017.
28. Lora Kolodny, "Impossible Foods CEO Pat Brown Says VCs Need to Ask Harder Scientific Questions," *TechCrunch*, May 22, 2017, https://techcrunch.com/2017/05/22/impossible-foods-ceo-pat-brown-says-vcs-need-to-ask-harder-scientific-questions/.
29. "A Proven Past, a Fortified Future," Dean Foods, http://www.deanfoods.com/our-company/about-us/brief-history.aspx, accessed November 12, 2017.
30. Kelley, "The Importance of Convenience in Consumer Purchasing," 32–38.
31. Emily Byrd, "2016: A Tipping Point for Food," Good Food Institute, December 20, 2016, http://www.gfi.org/2016-a-tipping-point-for-food.
32. Holzer, "Don't Put Vegetables in the Corner."
33. "Vegan Butcher Shop the Herbivorous Butcher Easily Surpasses Kickstarter Goal," *Fox 9*, November 25, 2014, http://www.fox9.com/news/1811266-story.
34. "Siblings Build a Butcher Shop for 'Meat'-Loving Vegans," NPR, December 7, 2014, http://www.npr.org/sections/thesalt/2014/12/07/369069078/siblings-build-a-butcher-shop-for-meat-loving-vegans; Melissa Locker, "A Vegan 'Butcher Shop' Is Opening in Minnesota," *Time*, January 7, 2016, http://time.com/4171727/a-vegan-butcher-shop-is-opening-in-minnesota/; Megan Charles, "Anyone for Tofu Turkey? 'Vegan Butcher' to Open in US," *Telegraph*, January 4, 2016, http://www.telegraph.co.uk/food-and-drink/news/anyone-for-tofu-turkey

-vegan-butcher-to-open-in-us/; Mahita Gajanan, "The Herbivorous Butcher: Sausage and Steak—but Hold the Slaughter," *Guardian*, January 29, 2016, https://www.theguardian.com/lifeandstyle/2016/jan/29/the-herbivorous-butcher-minneapolis-minnesota-vegan-meats; "Creole, Cold Cuts and Crepes," *Diners, Drive-Ins and Dives*, July 5, 2016, http://www.foodnetwork.com/shows/diners-drive-ins-and-dives/episodes/creole-cold-cuts-and-crepes.

5: The World's First Cultured Hamburger

1. Weather Underground, https://www.wunderground.com/history/airport/EHBK/2013/8/2/DailyHistory.html.
2. Conversation with Mark Post, 2017.
3. R.M. Hoffman, "The Beginning of Tissue Culture," Elsevier SciTech Connect, September 23, 2016, http://scitechconnect.elsevier.com/the-beginning-of-tissue-culture/.
4. Carrel, "On the Permanent Life of Tissues Outside of the Organism."
5. Ross, "The Smooth Muscle Cell II."
6. Leo Hickman, "Fake Meat: Burgers Grown in Beakers," *Wired*, July 31, 2009, http://www.wired.co.uk/article/fake-meat-burgers-grown-in-beakers; Michael Specter, "Test-Tube Burgers," *New Yorker*, May 23, 2011, http://www.newyorker.com/magazine/2011/05/23/test-tube-burgers.
7. Stephen Pincock, "Meat, In Vitro?," *Scientist*, September 1, 2007, http://www.the-scientist.com/?articles.view/articleNo/25358/title/Meat—in-vitro-/.
8. Ionat Zurr and Oron Catts, "I Feel Like Fake Frogs' Legs Tonight," *NY Arts*, June 25, 2006, http://www.nyartsmagazine.com/?p=3048; "Semi-Living Food: 'Disembodied Cuisine,'" Tissue Culture and Art Project, http://www.tca.uwa.edu.au/disembodied/dis.html.
9. Van Mensvoort and Grievink, *The In Vitro Meat Cookbook*.
10. Hickman, "Fake Meat."
11. "Very charismatic" is a quote from a conversation with Mark Post, 2017.
12. Dutch regulations on animal testing would have made it difficult to get live cow cells, so Post ended up taking cow cells from a slaughterhouse near his lab. Fortunately, this aspect doesn't seem to have led to much negative reception, and there's no reason to think that the cells couldn't come from live animals in the future. I hope future production uses live animals to avoid any unnecessary ethical concerns, although depending on the exact situation, keeping live animals might be more worrisome than taking scraps from a slaughterhouse.
13. Conversation with Mark Post, 2017.
14. "Burger Tasting London Aug. 2013," YouTube, https://www.youtube.com/watch?v=_Cy2x2QR968.
15. Conversation with Ryan Pandya, 2017.
16. I personally think that "lightbulb moments" often result from a gradual culmination of insights.
17. "Lab Meat: Tastes Like a Million Bucks," People for the Ethical

Treatment of Animals, http://www.peta.org/blog/lab-meat-tastes-like-million-bucks/.

18. "Clara Foods: Egg Whites Without Hens," New Harvest, http://www.new-harvest.org/clara_foods, accessed November 12, 2017.
19. Real Vegan Cheese, https://realvegancheese.org/; Biocurious, http://biocurious.org/.
20. "Sothic Bioscience: Protecting Human Lives While Preserving an Ancient Species," New Harvest, http://www.new-harvest.org/sothic-bioscience.
21. Teresa Novellino, "Modern Meadow Founder Andras Forgacs Makes Leather in a Brooklyn Lab," *New York Business Journal*, October 3, 2016, http://www.bizjournals.com/newyork/news/2016/10/03/modern-meadow-andras-forgacs-reinventor-upstart100.html; Karen Hao, "Would You Wear a Leather Jacket Grown in a Lab?" *Quartz*, February 5, 2017, https://qz.com/901643/would-you-wear-a-leather-jacket-grown-in-a-lab/.
22. Sarah Buhr, "Bolt Threads Debuts Its First Product, a $314 Tie Made from Spiderwebs," *TechCrunch*, March 10, 2017, https://techcrunch.com/2017/03/10/bolt-threads-debuts-its-first-product-a-314-tie-made-from-spiderwebs/.
23. Conversation with Uma Valeti, 2017.
24. Nicola Jones, "Meat-Growing Researcher Suspended," *Nature*, February 25, 2011, http://www.nature.com/news/2011/110225/full/news.2011.119.html; "Meat the Team," Memphis Meats, http://www.memphismeats.com/the-team/; Thomas Bailey Jr., "Futuristic Food Made in Labs Named Memphis Meats," *Commercial Appeal*, February 16, 2016, http://archive.commercialappeal.com/business/entrepreneurs/futuristic-cultured-meat-takes-on-the-memphis-brand-2be5b44c-94c7–6d89-e053–0100007f2094–369345301.html.
25. Jason Matheny had conversations with Josh Tetrick about cultured meat that helped spark Tetrick's original interest in the field. Conversation with Josh Balk and Josh Tetrick, 2017. Elaine Watson, "Hampton Creek to Enter Clean Meat Market in 2018: 'We Are Building a Multi-Species, Multi-Product Platform,'" *FoodNavigator-USA*, June 27, 2017, http://www.foodnavigator-usa.com/R-D/Hampton-Creek-to-enter-clean-meat-market-in-2018.
26. Conversation with Eitan Fischer, 2017.
27. Watson, "Hampton Creek to Enter Clean Meat Market in 2018."
28. Conversation with Josh Balk, 2017.
29. Dana Kessler, "Israel Goes Vegan," *Tablet*, November 28, 2012, http://www.tabletmag.com/jewish-life-and-religion/117687/israel-goes-vegan; "In the Land of Milk and Honey, Israelis Turn Vegan," Reuters, July 21, 2015, http://www.reuters.com/article/israel-food-vegan-idUSL5N0YW1L420150721.
30. Conversation with Chen Cohen, Shaked Regev, Yaron Bogin, and other Israeli advocates, 2017.
31. Conversation with Modern Agriculture Foundation representatives Shaked Regev and Yaron Bogin on behalf of Animal Charity Evaluators, 2016.
32. Elaine Watson, "SuperMeat Founder: 'The First Company That Gets to

Market with Cultured Meat That Is Cost Effective Is Going to Change the World,'" *FoodNavigator-USA*, July 20, 2016, http://www.foodnavigator-usa.com/R-D/SuperMeat-founder-on-why-cultured-meat-will-change-the-world.

33. Conversation with Finless Foods founder Mike Selden, 2017.
34. "History," New Harvest, http://www.new-harvest.org/history.
35. Scott Plous, "Re 'PETA's Latest Tactic: $1 Million for Fake Meat' (news article, April 21)," April 21, 2008, https://www.nytimes.com/2008/04/26/opinion/l26peta.html; Bruce Friedrich, "'Clean Meat': The 'Clean Energy' of Food," Good Food Institute, September 6, 2016, http://www.gfi.org/clean-meat-the-clean-energy-of-food.
36. Meera Zassenhaus, "On the Name 'Cultured Meat,'" New Harvest, July 7, 2016, https://medium.com/@NewHarvest/on-the-name-cultured-meat-1cc421544085.
37. Conversation with Mark Post, 2017; Jacob Bunge, "Cargill Invests in Startup That Grows 'Clean Meat' from Cells," *Wall Street Journal*, August 23, 2017, https://www.wsj.com/articles/cargill-backs-cell-culture-meat-1503486002.
38. Greig, "'Clean' Meat or 'Cultured' Meat."
39. Private communication with New Harvest staff, 2017.
40. Specht and Lagally, "Mapping Emerging Industries."
41. Social versus technological change is a complex topic with some important questions I left out, such as "Doesn't technological change lead to social change by reducing cognitive dissonance?" and "What makes whether or not animal farming ends so important, given that speeding it up still affects billions of animals?" and the more complicated "If humanity's values could become fixed at some point, such as due to an intelligence explosion or interstellar expansion, might speeding up the end of animal farming make a huge, directional difference?" For a full discussion of the arguments for a focus on social technology or social change, see Sentience Institute, "Social Change vs. Food Technology."

6: The Psychology of Animal-Free Food

1. Jesse Singal, "The 4 Ways People Rationalize Eating Meat," *New York*, June 4, 2015, http://nymag.com/scienceofus/2015/06/4-ways-people-rationalize-eating-meat.html.
2. Mark Joseph Stern, "A Little Guilt, a Lot of Energy Savings," *Slate*, March 1, 2013, http://www.slate.com/articles/technology/the_efficient_planet/2013/03/opower_using_smiley_faces_and_peer_pressure_to_save_the_planet.html; Hansen and Graham, "Preventing Alcohol, Marijuana, and Cigarette Use."
3. Salganik, Dodds, and Watts, "Experimental Study of Inequality and Unpredictability."
4. Saul McLeod, "Asch Experiment," SimplyPsychology, 2008, http://www.simplypsychology.org/asch-conformity.html; Sarah Knapton, "Nine in 10 People Would Electrocute Others If

Ordered, Rerun of Infamous Milgram Experiment Shows," *Telegraph*, March 14, 2017, http://www.telegraph.co.uk/science/2017/03/14/nine-10-people-would-electrocute-others-ordered-re-run-milgram/.

5. Joshua Barajas, "How the Nazi's Defense of 'Just Following Orders' Plays Out in the Mind," *NewsHour*, PBS, February 20, 2016, http://www.pbs.org/newshour/rundown/how-the-nazis-defense-of-just-following-orders-plays-out-in-the-mind/.
6. Jacy Reese, "Our Initial Thoughts on the Mercy For Animals Facebook Ads Study," Animal Charity Evaluators, February 19, 2016, https://animalcharityevaluators.org/blog/our-initial-thoughts-on-the-mfa-facebook-ads-study/.
7. Conversations with Chen Cohen and other Israeli advocates, 2016 and 2017.
8. Singal, "The 4 Ways People Rationalize Eating Meat."
9. "Meatless Meals: The Benefits of Eating Less Meat," Mayo Clinic, http://www.mayoclinic.org/healthy-lifestyle/nutrition-and-healthy-eating/in-depth/meatless-meals/art-20048193, accessed November 12, 2017; Roberto A. Ferdman, "Stop Eating So Much Meat, Top U.S. Nutritional Panel Says," *Washington Post*, February 19, 2015, https://www.washingtonpost.com/news/wonk/wp/2015/02/19/eating-a-lot-of-meat-is-hurting-the-environment-and-you-should-stop-top-u-s-nutritional-panel-says/; Tuso et al., "Nutritional Update for Physicians."
10. There is little evidence, and significant disagreement, on what exactly human ancestors ate, but scientists seem to agree that the amount of meat, dairy, and eggs was much less than that consumed today. See, for example, Rob Dunn, "Human Ancestors Were Nearly All Vegetarians," *Scientific American*, July 23, 2012, http://news.nationalgeographic.com/news/2005/02/0218_050218_human_diet.html.
11. Rozin, "The Meaning of 'Natural.'"
12. Macdonald and Vivalt, "Effective Strategies for Overcoming the Naturalistic Heuristic."
13. Dorea Reeser, "Natural Versus Synthetic Chemicals Is a Gray Matter," *Scientific American*, April 10, 2013, https://blogs.scientificamerican.com/guest-blog/natural-vs-synthetic-chemicals-is-a-gray-matter/.
14. The forty-two-days figure is from "The Life of: Broiler Chickens," Compassion in World Farming, https://www.ciwf.org.uk/media/5235306/The-life-of-Broiler-chickens.pdf, last modified January 5, 2013. Maximum lifespan isn't a popular or easy scientific research question, so I was unable to find a rigorous estimate, but the fifteen-years figure is my ballpark estimate based on my experience raising chickens with unusually good health care and my discussions with farmed animal experts. I think this figure could plausibly be as low as ten years or as high as twenty years. There are a few anecdotes of chickens living beyond twenty years, but that seems to be a rare exception.
15. Ilya Somin, "Over 80 Percent of Americans Support 'Mandatory Labels on Foods Containing DNA,'" *Washington Post*, January 17, 2015,

https://www.washingtonpost.com/news/volokh-conspiracy/wp/2015/01/17/over-80-percent-of-americans-support-mandatory-labels-on-foods-containing-dna/.

16. US Food and Drug Administration, "'Natural' on Food Labeling," https://www.fda.gov/Food/GuidanceRegulation/GuidanceDocumentsRegulatoryInformation/LabelingNutrition/ucm456090.htm accessed November 12, 2017.
17. Erin Brodwin, "This Cornell Scientist Saved an $11-Million Industry—and Ignited the GMO Wars," *Business Insider*, June 23, 2017, http://www.businessinsider.com/gmo-controversy-beginning-fruit-2017-6; Brad Plumer, "More Than 100 Nobel Laureates Are Calling on Greenpeace to End Its Anti-GMO Campaign," *Vox*, June 30, 2016, https://www.vox.com/2016/6/30/12066826/greenpeace-gmos-nobel-laureates; Federoff and Brown, *Mendel in the Kitchen*, 299–300.
18. Gabriel Rangel, "From Corgis to Corn: A Brief Look at the Long History of GMO Technology," Graduate School of Arts and Sciences blog, Harvard University, August 9, 2015, http://sitn.hms.harvard.edu/flash/2015/from-corgis-to-corn-a-brief-look-at-the-long-history-of-gmo-technology/.
19. Rochelle Kirkham, "Taste Test: Does the Future of Meat Lie in a Lab?," *Japan Times*, September 16, 2017, https://www.japantimes.co.jp/life/2017/09/16/food/taste-test-future-meat-lie-lab/; Yuki Hanyu, "Shojinmeat," presented at the International Conference on Cultured Meat, 2016, and New Harvest conference, 2017; Chase Purdy, "A Japanese Food Startup Is Giving High School Kids Meat-Growing Machines," *Quartz*, October 16, 2017, https://qz.com/1103280/japans-shojinmeat-project-takes-a-novel-approach-to-lab-grown-meat/.
20. Siegrist and Sütterlin, "Importance of Perceived Naturalness."
21. Chris Bryant, "Consumer Acceptance of Clean Meat: A Systematic Review," presented at the International Conference on Cultured Meat, Maastricht, September 2017.
22. Azagba and Sharaf, "The Effect of Graphic Cigarette Warning Labels."
23. For example, see the "Public Narrative" section in Mohorčich, "What Can Nuclear Power Teach Us."
24. Bob Fischer at Texas State University has made the moral case against "animal-friendly" animal products in his paper "You Can't Buy Your Way Out of Veganism."
25. Reese, "Even at Whole Foods' Best Farm, Turkeys Suffer."
26. Regarding fish, unfortunately the situation is so dire that there aren't even a meaningful number of fish farms that purport to use humane methods.
27. Daley et al., "A Review of Fatty Acid Profiles."
28. Jayson Lusk, "Why Industrial Farms Are Good for the Environment," *New York Times*, September 23, 2016, https://www.nytimes.com/2016/09/25/opinion/sunday/why-industrial-farms-are-good-for-the-environment.html.
29. Gidon Eshel, "Grass-Fed Beef Packs a Punch to Environment," Reuters, April 8, 2010, http://blogs.reuters.com/environment/2010/04/08/grass

-fed-beef-packs-a-punch-to-environment/; Capper, "Is the Grass Always Greener?"

30. There is also the consideration of economies of scale if specialty farms became more popular, though because most economies of scale come through industrialization at the cost of the intrinsic features of specialty farms, this effect seems minimal.
31. Witwicki, "Sentience Institute Global Farmed & Factory Farmed Animals Estimates"; Reese, "Sentience Institute US Factory Farming Estimates."
32. Reese, "Survey of US Attitudes Towards Animal Farming." Also consider that 2016 Nielsen data suggests specialty meat sales were tiny in comparison to conventional meat sales. For fresh meat, the conventional market was $31.9 billion while organic was $474.7 million (1.5 percent of conventional) and grass-fed was $288.2 million (0.9 percent). This doesn't account for chickens, who are more likely to live on factory farms and produce much less meat per animal than cows. Food Marketing Institute and North American Meat Institute, "The Power of Meat."
33. Jay Shooster, "Guest Column: Animal Welfare Cannot Be Dismissed," *Independent Florida Alligator*, December 2, 2015, http://www.alligator.org/opinion/columns/article_45dfb5b6-98b2-11e5-85f8-83c483e675c4.html.
34. Bastian et al., "Don't Mind Meat?"

7: Evidence-Based Social Change

1. In this section and throughout chapters 7 and 8, I will lean heavily on content I have already written that summarizes my research in the field. For this section, see Reese, "The Animal-Free Food Movement."
2. Faunalytics, "Study of Current and Former Vegetarians and Vegans."
3. Mercy For Animals, "Four Out of Five Americans Want Restaurants and Grocers to End Cruel Factory Farming Practices," PR Newswire, July 13, 2017, http://www.prnewswire.com/news-releases/four-out-of-five-americans-want-restaurants-and-grocers-to-end-cruel-factory-farming-practices-300487484.html.
4. Humane Society of the United States, "Initiative and Referendum History—Animal Protection Issues," accessed November 12, 2017, http://www.humanesociety.org/assets/pdfs/legislation/ballot_initiatives_chart.pdf.
5. Food Marketing Institute and North American Meat Institute, "The Power of Meat."
6. Reese, "Survey of US Attitudes Towards Animal Farming."
7. Crothers, "Free Produce Movement." The British antislavery movement is also useful evidence of the effectiveness of institutional change. See Witwicki, "Social Movement Lessons from the British Antislavery Movement."
8. George Monbiot, "We Cannot Change the World by Changing Our Buying Habits," *Guardian*, November 6, 2009, https://www.theguardian.com/environment/georgemonbiot/2009/nov/06/green-consumerism.
9. Mazar and Zhong, "Do Green Products Make Us Better People?"
10. Compassion in World Farming, a leading farmed animal protection

organization, published an interesting article on "What We Can Learn from the Anti-Smoking Movement," May 5, 2015, http://www.ciwf.org.uk/news/2015/05/what-we-can-learn-from-the-anti-smoking-movement-f1.

11. Cameron and Payne, "Escaping Affect."
12. Sethu, "How Many Animals Does a Vegetarian Save?"
13. Durkheim, *The Elementary Forms of the Religious Life*; Gabriel et al., "The Psychological Importance of Collective Assembly."
14. StreetAuthority, "How the 'Death of Meat' Could Impact Your Portfolio," Nasdaq, January 22, 2015, http://www.nasdaq.com/article/how-the-death-of-meat-could-impact-your-portfolio-cm435607; Caitlin Dewey, "Is This the Beginning of the End of Meat?," *Washington Post*, March 17, 2017, https://www.washingtonpost.com/news/wonk/wp/2017/03/17/is-this-the-beginning-of-the-end-of-meat/; Dan Murphy, "Meat of the Matter: Post-Animal Ag?," *Drovers*, September 20, 2016, https://web.archive.org/web/20160921201139/http://www.cattlenetwork.com/community/meat-matter-post-animal-ag.
15. Michael McCullough, "The Myth of Moral Outrage," Center for Humans and Nature, http://www.humansandnature.org/mind-morality-michael-mccullough, accessed November 12, 2017; Klandermans, van der Toorn, and van Stekelenburg, "Embeddedness and Identity."
16. Goodenough, "Moral Outrage."
17. Wakslak et al., "Moral Outrage Mediates the Dampening Effect of System Justification."
18. This argument can also be applied to celebrity endorsements, which might reach large numbers of people but also might make animal-free food seem more like a passing fad, similar to other things celebrities endorse, like the newest health craze or fashion trend.
19. "Meat Lovers Bite Back as Petition Calls for In-N-Out to Make Veggie Burger," *Fox News*, September 21, 2016, http://www.foxnews.com/leisure/2016/09/21/meat-lovers-bite-back-as-petition-calls-for-in-n-out-to-make-veggie-burger/.
20. Paluck and Ball, "Social Norms Marketing."
21. Bastian et al., "Don't Mind Meat?"
22. Reese, "Confrontation, Consumer Action, and Triggering Events."
23. Anderson, "Protection for the Powerless."
24. Animal Charity Evaluators, "Environmentalism."
25. Mohorčich, "What Can Nuclear Power Teach Us."
26. Bongiorno, Bain, and Haslam, "When Sex Doesn't Sell"; Wirtz, Sparks, and Zimbres, "The Effect of Exposure to Sexual Appeals"; Chen, "Booth Babes Don't Work."
27. Conversation with Alan Darer on behalf of Animal Charity Evaluators, 2017, https://animalcharityevaluators.org/research/charity-reviews/conversations/conversation-with-alan-darer/.
28. Eric Holthaus, "Stop Scaring People About Climate Change. It Doesn't Work," *Grist*, July 10, 2017, http://grist.org/climate-energy/stop-scaring-people-about-climate-change-it-doesnt-work/.
29. Jacy Reese, "Why Is It So Hard to Care About Large Groups of Animals?"

Dodo, January 26, 2016, https://www.thedodo.com/why-is-it-so-hard-to-care-about-large-groups-of-animals-1573188602.html.

30. "This Day in History: Baby Jessica Rescued from a Well as the World Watches," History.com, http://www.history.com/this-day-in-history/baby-jessica-rescued-from-a-well-as-the-world-watches, accessed November 12, 2017.
31. Desvousges et al., *Measuring Nonuse Damages.*
32. Reese, "Confrontation, Consumer Action, and Triggering Events."
33. Goodman, *Of One Blood*, 124.
34. Berger and Milkman, "What Makes Online Content Viral?"; Wakslak et al., "Moral Outrage Mediates the Dampening Effect of System Justification"; Klandermans, van der Toorn, and van Stekelenburg, "Embeddedness and Identity."
35. Nyhan and Reifler, "When Corrections Fail"; Haglin, "The Limitations of the Backfire Effect."

8: Broadening Horizons

1. Charles Tahler, "How Many Adults Are Vegetarian?," *Vegetarian Journal*, 2006, https://www.fda.gov/OHRMS/DOCKETS/98fr/FDA-1998-P-0032-bkg-reference01.pdf; Frank Newport, "In U.S., 5% Consider Themselves Vegetarians," Gallup, July 26, 2012, http://news.gallup.com/poll/156215/consider-themselves-vegetarians.aspx; Faunalytics, "Study of Current and Former Vegetarians and Vegans."
2. Sethu, "Meat Consumption Patterns by Race and Gender."
3. Nzinga Young, "Here's Why Black People Don't Go Vegan," *Huffington Post*, May 19, 2016, http://www.huffingtonpost.com/nzinga-young/heres-why-black-people-do_b_10028678.html; Alexandra Phanor-Faury, "Vegetarianism: A Black Choice," *Ebony*, April 8, 2015, http://www.ebony.com/life/vegetarianism-a-black-choice-333; Aph Ko, "3 Reasons Black Folks Don't Join the Animal Rights Movement—And Why We Should," *Everyday Feminism*, September 18, 2015, http://everydayfeminism.com/2015/09/black-folks-animal-rights-mvmt/.
4. Ko, "3 Reasons Black Folks Don't Join the Animal Rights Movement—And Why We Should."
5. "Feed: Meat Ratios," A Well-Fed World, October 26, 2015, http://awfw.org/feed-ratios/.
6. Haidt, *The Righteous Mind.*
7. Bain et al., "Promoting Pro-Environmental Action in Climate Change Deniers."
8. Feygina, Jost, and Goldsmith, "System Justification"; Dhonta and Hodson, "Why Do Right-Wing Adherents Engage."
9. Aryeh Neier, "Brown v. Board of Ed: Key Cold War Weapon," Reuters, May 14, 2014, http://blogs.reuters.com/great-debate/2014/05/14/brown-v-board-of-ed-key-cold-war-weapon/.
10. Matt Frazier and Stepfanie Romine, "How to Be an Athlete on a Plant-Based Diet," *Sports Illustrated*, May 18, 2017, https://www.si.com/edge/2017/05/18/no-meat-athlete-cookbook-plant-based-diet.

11. Toni Okamoto, "Plant Based on a Budget," http://www.toniokamoto.com/#/plantbasedonabudget/, accessed November 12, 2017.
12. For an example of strongly advocating this idea, see Nick Cooney, "The 2012 Presidential Election and the Future of Veg Advocacy," December 11, 2012, https://ccc.farmsanctuary.org/the-2012-presidential-election-and-the-future-of-veg-advocacy/. Note that a focus on young people is more tenable than other demographic focuses because young people have a greater influence on future social and policy change.
13. Bollard, "Farm Animal Statistics."
14. Pi, Rou, and Horowitz, "Fair or Fowl?"
15. Mercy For Animals, "Four out of Five Americans"; You et al., "A Survey of Chinese Citizens' Perceptions on Farm Animal Welfare."
16. Bollard, "How Can We Improve Farm Animal Welfare in India"; US Department of Agriculture, Economic Research Service, *US Per Capita Egg Consumption 1950–2008*, http://www.humanesociety.org/assets/pdfs/farm/Per-Cap-Cons-Eggs-1.pdf, accessed November 12, 2017.
17. India Parliament, *The Prevention of Cruelty to Animals Act, 1960*, http://www.moef.nic.in/sites/default/files/No.59.pdf; "Hope for Hens: India Agrees That Battery Cages Are Illegal," Humane Society International, May 13, 2013, http://www.hsi.org/world/india/news/news/2013/05/victory_hens_india_051413.html; Bollard, "How Can We Improve Farm Animal Welfare in India."
18. Bollard, "Farm Animal Statistics."
19. Devon Haynie, "These Are the World's Most Influential Countries," *US News and World Report*, March 7, 2017, https://www.usnews.com/news/best-countries/best-international-influence.
20. Christine Mahoney, "Why Lobbying in America Is Different," *Politico*, June 4, 2009, last modified April 12, 2014, http://www.politico.eu/article/why-lobbying-in-america-is-different/.
21. Environmental Performance Index, "Country Rankings"; World Animal Protection, *Animal Protection Index*; Tandon et al., "Measuring Overall Health"; US Central Intelligence Agency, *The World Factbook: Obesity—Adult Prevalence Rate*, https://www.cia.gov/library/publications/the-world-factbook/fields/2228.html, accessed November 12, 2017; International Diabetes Federation, *IDF Diabetes Atlas*.
22. "Each Country's Share of CO2 Emissions," Union of Concerned Scientists, last modified November 18, 2014, http://www.ucsusa.org/global_warming/science_and_impacts/science/each-countrys-share-of-co2.html; Christina Nunez, "China Poised for Leadership on Climate Change After U.S. Reversal," *National Geographic*, March 28, 2017, http://news.nationalgeographic.com/2017/03/china-takes-leadership-climate-change-trump-clean-power-plan-paris-agreement/.
23. Oliver Milman and Stuart Leavenworth, "China's Plan to Cut Meat Consumption by 50% Cheered by Climate Campaigners," *Guardian*, June 20, 2016, https://www.theguardian.com/world/2016/jun/20/chinas-meat-consumption-climate-change.
24. "Making a Difference for Pigs in China," Compassion in World

Farming, September 25, 2014, https://www.ciwf.org.uk/news/2014/09/making-a-difference-for-pigs-in-china.

25. Edward Wong, "Clampdown in China Restricts 7,000 Foreign Organizations," *New York Times*, April 28, 2016, https://www.nytimes.com/2016/04/29/world/asia/china-foreign-ngo-law.html.
26. James Nason, "Temple Grandin's 'Thumbs Up' for Japfa's Chinese Feedlot," *Beef Central*, July 9, 2014, http://www.beefcentral.com/live-export/temple-grandins-thumbs-up-for-japfas-chinese-feedlot/; "McDonald's to Boost China Supplier Audits After Food Safety Scandal," Reuters, September 2, 2014, http://www.reuters.com/article/us-mcdonalds-china-idUSKBN0GX0MW20140902.
27. Nick Pachelli, "The Road to a Post-Meat World Starts in China," *Vice*, October 7, 2016, https://munchies.vice.com/en_us/article/the-road-to-a-post-meat-world-starts-in-china.
28. Michael Charles Tobias, "Animal Rights in China," *Forbes*, November 2, 2012, https://www.forbes.com/sites/michaeltobias/2012/11/02/animal-rights-in-china/.
29. Bollard, "Farm Animal Statistics."
30. Drake Baer, "This Startup Is Using Machine Learning to Create Animal Product Substitutes," *Business Insider*, April 26, 2016, http://www.businessinsider.com/chilean-startup-uses-machine-learning-for-meat-subsitutes-2016–4.
31. "Animal Equality's Work in Latin America," Animal Equality, February 16, 2016, http://www.animalequality.net/node/857.
32. Elizabeth McSheffrey and Jenny Uechi, "Meet the Vegan Saudi Prince Who's Turning the Lights On in Jordan," *National Observer*, February 10, 2017, http://www.nationalobserver.com/2017/02/10/news/meet-vegan-saudi-prince-whos-turning-lights-jordan; "Prince Al-Waleed Bin Talal bin Abdulaziz al Saud," *Forbes*, https://www.forbes.com/profile/prince-alwaleed-bin-talal-alsaud/, accessed November 12, 2017.

9: The Expanding Moral Circle, Revisited

1. Witwicki, "Sentience Institute Global Farmed & Factory Farmed Animals Estimates."
2. Gerland et al., "World Population Stabilization Unlikely This Century."
3. Sarah Fecht, "Elon Musk Wants to Put Humans on Mars by 2025," *Popular Science*, June 2, 2016, http://www.popsci.com/elon-musk-wants-to-put-humans-on-mars-by-2025.
4. Bostrom, "Astronomical Waste"; Spurio, *Particles and Astrophysics*, 207.
5. The topic of artificial intelligence safety is discussed at length in Bostrom, *Superintelligence*.
6. See comments on "Trapped Mouflon (Wild Sheep) Rescued by Jogger," YouTube, March 30, 2015, https://www.youtube.com/watch?v=0kmWsd_wMeY.
7. This is actually being done by Patrick Kilonzo Mwalua as reported in global media outlets in February 2017. See Christian Cotroneo, "Man

Drives Hours Every Day in Drought to Bring Water to Wild Animals," *Dodo*, February 17, 2017, https://www.thedodo.com/water-man-kenya-animals-2263728686.html.

8. Wild animals are currently outside of humanity's moral circle in the dimension of the cause of their suffering (natural rather than anthropogenic) and the dimension of species (nonhuman and often invertebrate rather than human and vertebrate) as discussed in this section. However, they are also an important example of the "act versus omission" dimension. We see clearer harm when it's caused by an act, such as eating animal-based food, rather than when it's caused by an omission, such as not intervening in the wild to provide vaccines. For more reading on the issue of wild animal suffering, the canonical text is Brian Tomasik, "The Importance of Wild-Animal Suffering," Foundational Research Institute, July 2009, https://foundational-research.org/the-importance-of-wild-animal-suffering/.
9. Webb, "Cognition in Insects"; Klein and Barron, "Insects Have the Capacity for Subjective Experience."
10. Tracey et al., "Painless."
11. Bateson et al., "Agitated Honeybees Exhibit Pessimistic Cognitive Biases."
12. Bug estimate from Tomasik, "How Many Wild Animals Are There?" Farmed animal estimate, 209.6 billion, from Witwicki, "Global Farmed & Factory Farmed Animals Estimates." Result is 4.77 to 47.7, rounded for simplicity and to avoid giving false precision.
13. This possibility of uploading human brains into the digital world is explored at length in Hanson, *The Age of Em*.
14. Michio Kaku, "Your Cell Phone Has More Computing Power Than NASA Circa 1969," Knopf Doubleday Publishing Group, from Kaku, *Physics of the Future* (New York: Doubleday, 2011), http://knopfdoubleday.com/2011/03/14/your-cell-phone/, accessed November 12, 2017.
15. "Carpageddon: Australia Plans to Kill Carp with Herpes," *BBC News*, May 3, 2016, http://www.bbc.com/news/world-australia-36189409.
16. In addition to how different advocacy framings affect moral circle expansion on their own, we should also consider that if effectiveness-focused advocates emphasize environmental and health arguments for animal-free food and the less effectiveness-focused advocates emphasize animal welfare, the reputation of animal welfare could decrease because of the tactics of those advocates, exacerbating the curtailment of future progress.
17. Grace et al., "When Will AI Exceed Human Performance?"
18. All the predictions in this section assume that human civilization will continue its basic current trajectory, e.g., we won't go extinct due to nuclear war, we won't develop an authoritarian world government that impedes social and scientific progress.
19. Conversation with Houman Shahi, 2017.
20. This is an 80 percent prediction interval, meaning that I think it's 20 percent likely that this stage will arrive sooner than ten years or later than thirty years. I want to put forward a specific timeline for the purposes of being clear about my views and making them testable, but I also want to

clarify that these numbers rely heavily on intuition and are not intended to suggest precision or confidence.

21. Witwicki, "Social Movement Lessons from the British Antislavery Movement."
22. *Merriam-Webster's Collegiate Dictionary*, s.v. "weltschmerz."
23. Good Food Institute, "White Space Company Ideas."

BIBLIOGRAPHY

Albala, Ken. *Food: A Cultural Culinary History*. Audio recording. Chantilly, VA: Great Courses, 2013.

American Egg Board. *2015 Annual Report*. http://www.aeb.org/images/PDFs/AboutAEB/AEB_Annual_Report_2015.pdf.

American Humane Association. "2014 Humane Heartland Farm Animal Welfare Survey," http://www.americanhumane.org/app/uploads/2016/08/2014-humane-heartland-farm-survey.pdf.

Anderson, Jerry L. "Protection for the Powerless: Political Economy History Lessons for the Animal Welfare Movement." *Stanford Journal of Animal Law and Policy* 4, no. 1 (2011). http://escholarshare.drake.edu/bitstream/handle/2092/1476/SJALP_Anderson_final%20formatted_12%2020%20 2010%5B1%5D.pdf.

Animal Charity Evaluators. "Environmentalism." Accessed November 12, 2017. https://animalcharityevaluators.org/research/social-movement-analysis/environmentalism/.

———. "Exploratory Review: People for the Ethical Treatment of Animals (PETA)." https://animalcharityevaluators.org/research/charity-review/people-for-the-ethical-treatment-of-animals-peta/. Accessed November 9, 2017.

———. "2015 Comprehensive Review: Mercy For Animals." https://animalcharityevaluators.org/research/charity-review/mercy-for-animals/2015-dec/#comprehensive-review. Accessed November 9, 2017.

———. "Why Farmed Animals?" http://www.animalcharityevaluators.org/research/foundational-research/number-of-animals-vs-amount-of-donations/. Accessed November 9, 2017.

Aristotle. *History of Animals*. 350 BC. Translated by D'Arcy Wentworth Thompson. Accessed November 12, 2017. http://classics.mit.edu/Aristotle/history_anim.html.

Arndt, Milton. *Battery Brooding; a Complete Exposition of the Important Facts Concerning the Successful Operation and Handling of the Various Types of Battery Brooders*. New York: Orange Judd, 1931.

Azagba, Sunday, and Mesbah F. Sharaf. "The Effect of Graphic Cigarette Warning Labels on Smoking Behavior: Evidence from the Canadian Experience." *Nicotine & Tobacco Research* 15, no. 3 (2013): 708–17.

https://academic.oup.com/ntr/article/15/3/708/1091051/The-Effect-of-Graphic-Cigarette-Warning-Labels-on.

Bain, Paul G., et al. "Promoting Pro-environmental Action in Climate Change Deniers." *Nature Climate Change* 2, no. 8 (2012): 603. https://climateaccess.org/system/files/Bain_Promoting%20pro-environmental%20action.pdf.

Bastian, Brock, et al. "Don't Mind Meat? The Denial of Mind to Animals Used for Human Consumption." *Personality and Social Psychology Bulletin* 38, no. 2 (2012): 247–56. https://www.ncbi.nlm.nih.gov/pubmed/21980158.

Bastide, Nadia M., Fabrice H. F. Pierre, and Denis E. Corpet. "Heme Iron from Meat and Risk of Colorectal Cancer: A Meta-analysis and a Review of the Mechanisms Involved." *Cancer Prevention Research* 4, no. 2 (2011): 177–84. https://www.ncbi.nlm.nih.gov/pubmed/21209396.

Bateson, Melissa, et al. "Agitated Honeybees Exhibit Pessimistic Cognitive Biases." *Current Biology* 21, no. 12 (2011): 1070–73. https://www.ncbi.nlm.nih.gov/pmc/articles/PMC3158593/.

Benson, Tess. "Advancing Aquaculture: Fish Welfare at Slaughter." Humane Slaughter Association. http://seafood.oregonstate.edu/.pdf%20Links/Fish%20Welfare%20at%20Slaughter%20by%20Tess%20Benson%20-%20Winston%20Churchill%20Memorial%20Trust.pdf. Accessed November 12, 2017.

Berger, Jonah, and Katherine L. Milkman. "What Makes Online Content Viral?" *Journal of Marketing Research* 49, no. 2 (2012): 192–205. http://jonahberger.com/wp-content/uploads/2013/02/ViralityB.pdf.

Berndorff, Jan. "Klug, aber schweinisch." *Bild der Wissenschaft* 3 (2016): 11–17. http://www.wissenschaft.de/heft_inhaltsverzeichnis/-/journal_content/56/12054/10118383.

Blokhuis, H. J., and J. G. Arkes. "Some Observations on the Development of Feather-Pecking in Poultry." *Applied Animal Behaviour Science* 12, no. 1–2 (1984): 145–57. http://www.sciencedirect.com/science/article/pii/0168159184901047.

Bollard, Lewis. "Farm Animal Statistics." https://docs.google.com/spreadsheets/d/1uF3x_DuG13V6NpkP4DQyFutZI6Pwppy5R6qGArPZ1ho/. Accessed November 12, 2017.

———. "How Can We Improve Farm Animal Welfare in India?" http://us14.campaign-archive1.com/?u=66df320da8400b581cbc1b539&id=1096590fac&e=b669f8c82c. Accessed November 12, 2017.

———. "Initial Grants to Support Corporate Cage-free Reforms." Open Philanthropy Project. March 31, 2016. http://www.openphilanthropy.org/blog/initial-grants-support-corporate-cage-free-reforms.

Bongiorno, Renata, Paul G. Bain, and Nick Haslam. "When Sex Doesn't Sell: Using Sexualized Images of Women Reduces Support for Ethical Campaigns." *PLOS ONE* 8, no. 12 (2013). http://journals.plos.org/plosone/article?id=10.1371%2Fjournal.pone.0083311.

Bostrom, Nick. "Astronomical Waste." *Utilitas* 15, no. 3 (2003): 308–14. http://www.nickbostrom.com/astronomical/waste.html.

———. *Superintelligence: Paths, Dangers, Strategies.* New York: Oxford University Press, 2014.

Brewer, M. Susan. "The Chemistry of Beef Flavor: Executive Summary." National Cattlemen's Beef Association. http://www.beefissuesquarterly.com/CMDocs/BeefResearch/PE_Executive_Summaries/The_Chemistry_of_Beef_Flavor.pdf.

Cameron, C. Daryl, and B. Keith Payne. "Escaping Affect: How Motivated Emotion Regulation Creates Insensitivity to Mass Suffering." *Journal of Personality and Social Psychology* 100, no. 1 (2011): 1–15. http://www.ncbi.nlm.nih.gov/pubmed/21219076.

Capper, Judith L. "Is the Grass Always Greener? Comparing the Environmental Impact of Conventional, Natural and Grass-Fed Beef Production Systems." *Animals* 2, no. 2 (2012): 127–43. http://www.mdpi.com/2076–2615/2/2/127.

Carrel, Alexis. "On the Permanent Life of Tissues Outside of the Organism." *Journal of Experimental Medicine* 15, no. 5 (1912): 516–28. http://jem.rupress.org/content/jem/15/5/516.full.pdf.

Cooney, Nick. "The 2012 Presidential Election and the Future of Veg Advocacy." Farm Sanctuary, December 11, 2012. https://ccc.farmsanctuary.org/the-2012-presidential-election-and-the-future-of-veg-advocacy/.

Cottingham, John. "'A Brute to the Brutes?': Descartes' Treatment of Animals." *Philosophy* 53, no. 206 (1978): 551–59. http://people.whitman.edu/~herbrawt/classes/339/Descartes.pdf.

Crothers, A. Glenn. "Free Produce Movement." In *Encyclopedia of Antislavery and Abolition*, vol. 1. Edited by Peter P. Hinks and John R. McKivigan. Westport, CT: Greenwood, 2006.

Daley, Cynthia A., et al. "A Review of Fatty Acid Profiles and Antioxidant Content in Grass-Fed and Grain-Fed Beef." *Nutrition Journal* (2010): 9–10. https://www.ncbi.nlm.nih.gov/pmc/articles/PMC2846864/.

Darwin, Charles. *The Expression of the Emotions in Man and Animals.* London: John Murray, 1872. http://gruberpeplab.com/teaching/psych131_summer2013/documents/Lecture3_Darwin1872_EmotionalExpressioninManandAnimals.pdf.

———. *The Formation of Vegetable Mould Through the Actions of Earth Worms, with Observations on Their Habits.* London: John Murray, 1881.

Desvousges, William H., et al. *Measuring Nonuse Damages Using Contingent Valuation: An Experimental Evaluation of Accuracy*, 2nd ed. Research Triangle Park, NC: RTI Press, 2010. http://www.rti.org/sites/default/files/resources/bk-0001–1009_web.pdf.

De Waal, Frans. *Are We Smart Enough to Know How Smart Animals Are?* New York: W. W. Norton, 2016.

Dhonta, Kristof, and Gordon Hodson. "Why Do Right-Wing Adherents Engage in More Animal Exploitation and Meat Consumption?" *Personality and Individual Differences* 64 (2014): 12–17. http://www.sciencedirect.com/science/article/pii/S0191886914000944.

Dio, Cassius. *Roman History*. Translated by Earnest Cary. Accessed

November 12, 2017. http://penelope.uchicago.edu/Thayer/E/Roman/Texts/Cassius_Dio/66*.html#25.

Durkheim, Émile. *The Elementary Forms of the Religious Life*. Translated by Karen E. Fields. New York: Free Press, 1995.

Environmental Performance Index. "Country Rankings." http://epi.yale.edu/country-rankings. Accessed November 12, 2017.

Faunalytics. "Study of Current and Former Vegetarians and Vegans." December 2014. https://faunalytics.org/wp-content/uploads/2015/06/Faunalytics_Current-Former-Vegetarians_Tables-Methodology.pdf.

Federoff, Nina, and Nancy Marie Brown. *Mendel in the Kitchen*. Washington, DC: Joseph Henry Press, 2004.

Feygina, Irina, John T. Jost, and Rachel E. Goldsmith. "System Justification, the Denial of Global Warming, and the Possibility of 'System-Sanctioned Change.'" *Personality and Social Psychology Bulletin* 36, no. 3 (2010): 326–38. http://journals.sagepub.com/doi/abs/10.1177/0146167209351435.

Fischer, Bob. "You Can't Buy Your Way Out of Veganism." *Between the Species* 19, no. 1 (2016). http://digitalcommons.calpoly.edu/bts/vol19/iss1/8/.

Food Marketing Institute and North American Meat Institute. "The Power of Meat." 2017. http://www.meatconference.com/sites/default/files/books/Power_of_meat_2017.pdf.

Gabler, Neal. *Walt Disney: The Triumph of the American Imagination*. New York: Vintage Books, 2007.

Gabriel, Shira, et al. "The Psychological Importance of Collective Assembly: Development and Validation of the Tendency for Effervescent Assembly Measure (TEAM)." *Psychological Assessment* (2017). https://www.ncbi.nlm.nih.gov/pubmed/28263640.

Garrison, William Lloyd. *Thoughts on African Colonization*. Boston: Garrison and Knapp, 1832.

Gerland, Patrick, et al. "World Population Stabilization Unlikely This Century." *Science* 346, no. 6206 (2014): 234–37. http://science.sciencemag.org/content/early/2014/09/17/science.1257469.

Goodenough, Ward H. "Moral Outrage: Territoriality in Human Guise." *Zygon* 32, no. 1 (1997): 5–27. http://www.sscnet.ucla.edu/anthro/faculty/fiske/RM_PDFs/Goodenough_Moral_Territoriality.pdf.

Good Food Institute. "White Space Company Ideas." https://docs.google.com/document/d/1zCwLkwqwYzfzxwIm1-iHrvheRhMhbLYBxqEiz_7bHdE/. Accessed November 12, 2017.

Goodman, Paul. *Of One Blood: Abolitionism and the Origins of Racial Equality*. Berkeley: University of California Press, 1998.

Grace, Katja, et al. "When Will AI Exceed Human Performance? Evidence from AI Experts." Future of Humanity Institute, University of Oxford, May 30, 2017. https://arxiv.org/pdf/1705.08807.pdf.

Greig, Kieran. "'Clean' Meat or 'Cultured' Meat: A Randomized Trial Evaluating the Impact on Self-Reported Purchasing Preferences." Animal Charity Evaluators. May 10, 2017. https://animalcharityevaluators.org/blog/clean-meat-or-cultured-meat-a-randomized-trial-evaluating-the-impact-on-self-reported-purchasing-preferences/.

Haglin, Kathryn. "The Limitations of the Backfire Effect." *Research and Politics* (2017). http://journals.sagepub.com/doi/full/10.1177/2053168017716547.

Haidt, Jonathan. *The Righteous Mind: Why Good People Are Divided by Politics and Religion*. New York: Pantheon, 2012.

Hansen, William B., and John W. Graham. "Preventing Alcohol, Marijuana, and Cigarette Use Among Adolescents: Peer Pressure Resistance Training Versus Establishing Conservative Norms." *Preventative Medicine* 20, no. 3 (1991): 414–30. https://www.ncbi.nlm.nih.gov/pubmed/1862062.

Hanson, Robin. *The Age of Em: Work, Love, and Life When Robots Rule the Earth*. New York: Oxford University Press, 2016.

Herzog, Hal. *Some We Love, Some We Hate, Some We Eat: Why It's So Hard to Think Straight About Animals*. New York: Harper Perennial, 2010.

Holzer, Jillian. "Don't Put Vegetables in the Corner: Q&A with Behavioral Science Researcher Linda Bacon." World Resources Institute, June 12, 2017. http://www.wri.org/blog/2017/06/dont-put-vegetables-corner-qa-behavioral-science-researcher-linda-bacon.

Humane Society of the United States. "Farm Animal Statistics: Slaughter Totals." http://www.humanesociety.org/news/resources/research/stats_slaughter_totals.html. Accessed November 12, 2017.

———. "Welfare Issues with Gestation Crates for Pregnant Sows." February 2013. http://www.humanesociety.org/assets/pdfs/farm/HSUS-Report-on-Gestation-Crates-for-Pregnant-Sows.pdf.

———. "Welfare Issues with Selective Breeding of Egg-Laying Hens for Productivity." January 2007. http://www.fao.org/fileadmin/user_upload/animalwelfare/HSUS—Welfare%20Issues%20with%20Selective%20Breeding%20of%20Egg-Laying%20Hens%20for%20Productivity.pdf.

Institute of Medicine. *Dietary Reference Intakes for Energy, Carbohydrate, Fiber, Fat, Fatty Acids, Cholesterol, Protein, and Amino Acids*. Washington, DC: National Academies Press, 2005. https://www.nap.edu/catalog/10490/dietary-reference-intakes-for-energy-carbohydrate-fiber-fat-fatty-acids-cholesterol-protein-and-amino-acids.

International Diabetes Federation. *IDF Diabetes Atlas*, 8th ed. http://www.diabetesatlas.org/. Accessed November 12, 2017.

Jörgensen, Svea. "Empathy, Altruism and the African Elephant." Bachelor's thesis, Swedish University of Agricultural Sciences, 2015. http://stud.epsilon.slu.se/7913/11/jorgensen_s_150507.pdf.

Kellert, Stephen R. "American Attitudes Toward and Knowledge of Animals: An Update." *Advances in Animal Welfare Science* 85 (1984): 177–213. http://animalstudiesrepository.org/cgi/viewcontent.cgi?article=1000&context=acwp_sata.

Kelley, Eugene J. "The Importance of Convenience in Consumer Purchasing." *Journal of Marketing* 23, no. 1 (1958): 32–38. https://www.jstor.org/stable/1248014.

Klandermans, Bert, Jojanneke van der Toorn, and Jacquelien van Stekelenburg. "Embeddedness and Identity: How Immigrants Turn Grievances into Action." *American Sociological Review* 73, no. 6 (2008): 992–1012.

http://www.academia.edu/988489/Embeddedness_and_identity_How_immigrants_turn_grievances_into_action.

Klein, Colin, and Andrew B. Barron. "Insects Have the Capacity for Subjective Experience." *Animal Sentience* 100 (2016). http://animalstudiesrepository.org/cgi/viewcontent.cgi?article=1113&context=animsent.

Knoll, Laura P. "Origins of the Regulation of Raw Milk Cheeses in the United States." Third-Year Paper, Harvard Law School, April 26, 2005. https://dash.harvard.edu/bitstream/handle/1/8852188/Knoll05.pdf.

Krintirasa, Georgios A., et al. "On the Use of the Couette Cell Technology for Large Scale Production of Textured Soy-Based Meat Replacers." *Journal of Food Engineering* 169 (2016): 205–13. http://www.sciencedirect.com/science/article/pii/S026087741500374X.

Lebergott, Stanley. "Labor Force and Employment, 1800–1960." In *Output, Employment, and Productivity in the United States After 1800*, edited by Dorothy S. Brady, 117–204. http://www.nber.org/chapters/c1567.pdf.

Low, Phillip. "Cambridge Declaration on Consciousness." July 7, 2012. http://fcmconference.org/img/CambridgeDeclarationOnConsciousness.pdf.

Macdonald, Bobbie N. J., and Eva Vivalt. "Effective Strategies for Overcoming the Naturalistic Heuristic." August 11, 2017. https://osf.io/cb589/.

Marino, Lori, and Christina M. Colvin. "Thinking Pigs: A Comparative Review of Cognition, Emotion, and Personality in Sus domesticus." *International Journal of Comparative Psychology* 28 (2015). http://animalstudiesrepository.org/cgi/viewcontent.cgi?article=1042&context=acwp_asie.

Masson, Jeffrey Moussaieff. *When Elephants Weep: The Emotional Lives of Animals*. London: Vintage, 1996.

Mazar, Nina, and Chen-Bo Zhong. "Do Green Products Make Us Better People?" *Psychological Science* 21 (2010): 494–98. http://journals.sagepub.com/doi/abs/10.1177/0956797610363538.

Mercy For Animals. "Why We Work for Policy Change." http://www.mercyforanimals.org/why-policy-change. Accessed November 9, 2017.

Mohorčich, J., "What Can Nuclear Power Teach Us About the Institutional Adoption of Clean Meat?" Sentience Institute, November 27, 2017. http://www.sentienceinstitute.org/what-can-nuclear-energy-teach-us/.

Moore, Mario. "Food Labeling Regulation: A Historical and Comparative Survey." Third Year Paper, Harvard University, 2001. https://dash.harvard.edu/bitstream/handle/1/8965597/moorem.pdf?sequence=1.

Mosko, Mark S. *Quadripartite Structures: Categories, Relations and Homologies in Bush Mekeo Culture*. Cambridge, UK: Cambridge University Press, 1985.

Mullally, Conner, and Jayson L. Lusk. "The Impact of Farm Animal Housing Restrictions on Egg Prices, Consumer Welfare, and Production in California," *American Journal of Agricultural Economics* (2017). https://academic.oup.com/ajae/advance-article/doi/10.1093/ajae/aax049/4157679.

Nyhan, Brendan, and Jason Reifler. "When Corrections Fail: The Persistence of Political Misperceptions." *Political Behavior* 32, no. 2 (2010): 303–30. https://www.dartmouth.edu/~nyhan/nyhan-reifler.pdf.

Obee, Bruce, and Graeme Ellis. *Guardians of the Whales: The Quest to Study Whales in the Wild*. Anchorage: Alaska Northwest Books, 1992.
Paluck, Elizabeth Levy, and Laurie Ball. "Social Norms Marketing Aimed at Gender-Based Violence: A Literature Review and Critical Assessment." International Rescue Committee, 2010. https://static.squarespace.com/static/5186d08fe4b065e39b45b91e/t/52d1f24ce4b07fea759e4446/1389490764065/Paluck%20Ball%20IRC%20Social%20Norms%20Marketing%20Long.pdf.
Pasiakos, Stefan M., Sanjiv Agarwal, Harris R. Lieberman, and Victor L. Fulgoni III. "Sources and Amounts of Animal, Dairy, and Plant Protein Intake of US Adults in 2007–2010." *Nutrients* 7, no. 8 (2015): 7058–69. https://www.ncbi.nlm.nih.gov/pmc/articles/PMC4555161/.
Perry, Graham C. "Cannibalism." In *Welfare of the Laying Hen*. Edited by G. C. Perry. Wallingford, UK: CABI, 2004.
Pharr, Clyde. *The Theodosian Code and Novels, and the Sirmondian Constitutions*. Princeton, NJ: Princeton University Press, 1952.
Pi, Chendong, Zhang Rou, and Sarah Horowitz. "Fair or Fowl?: Industrialization of Poultry Production in China." Institute for Agriculture and Trade Policy. February 2014. https://www.iatp.org/sites/default/files/2017–05/2017_05_03_PoultryReport_f_web.pdf.
Pinker, Steven. *The Better Angels of Our Nature: Why Violence Has Declined*. New York: Viking, 2011.
Reese, Jacy. "The Animal-Free Food Movement Should Move Towards an Institutional Message." *Medium*. October 20, 2016. https://medium.com/@jacyreese/the-animal-free-food-movement-should-move-towards-an-institutional-message-534d7cd0298e.
———. "Confrontation, Consumer Action, and Triggering Events."Animal Charity Evaluators. July 14, 2015. https://animalcharityevaluators.org/blog/confrontation-consumer-action/.
———. "Sentience Institute US Factory Farming Estimates." Sentience Institute. https://docs.google.com/spreadsheets/d/1iUpRFOPmAE5IO4hO4PyS4MP_kHzkuM_-soqAyVNQcJc/. Accessed November 9, 2017.
———. "Survey of US Attitudes Towards Animal Farming and Animal-Free Food October 2017." Sentience Institute. https://www.sentienceinstitute.org/animal-farming-attitudes-survey-2017. Accessed November 20, 2017.
Robbins, J. A., et al. "Awareness of Ag-Gag Laws Erodes Trust in Farmers and Increases Support for Animal Welfare Regulations." *Food Policy* 61 (2016): 121–25. https://www.researchgate.net/profile/Becca_Franks/publication/297729566_Awareness_of_ag-gag_laws_erodes_trust_in_farmers_and_increases_support_for_animal_welfare_regulations/links/5728cf8008aef7c7e2c0c7ce.pdf.
Ross, Russell. "The Smooth Muscle Cell II: Growth of Smooth Muscle in Culture and Formation of Elastic Fibers." *Journal of Cell Biology* 50, no. 1 (1971): 172–86. https://www.ncbi.nlm.nih.gov/pmc/articles/PMC2108435/.
Rothgerber, Hank, and Frances Mican. "Childhood Pet Ownership,

Attachment to Pets, and Subsequent Meat Avoidance: The Mediating Role of Empathy Toward Animals." *Appetite* 79 (2014): 11–17. http://vegstudies.univie.ac.at/fileadmin/user_upload/p_foodethik/Rothgerber__Hank_2014._Childhood_pet_ownership__attachment_to_pets__and_subsequent_meat_avoidance.pdf.

Rozin, Paul. "The Meaning of 'Natural': Process More Important Than Content." *American Psychological Society* 16, no. 8 (2005): 652–58. https://www.ncbi.nlm.nih.gov/pubmed/16102069.

Sacks, Oliver. *An Anthropologist on Mars: Seven Paradoxical Tales.* New York: Knopf, 1995.

Safina, Carl. *Beyond Words: What Animals Think and Feel.* New York: Henry Holt, 2015.

Salganik, Matthew J., Peter Sheridan Dodds, and Duncan J. Watts. "Experimental Study of Inequality and Unpredictability in an Artificial Cultural Market." *Science* 311 (2006): 854–56. https://www.princeton.edu/~mjs3/salganik_dodds_watts06_full.pdf.

Sato, Nobuya, Ling Tan, Kazushi Tate, and Maya Okada. "Rats Demonstrate Helping Behavior Toward a Soaked Conspecific." *Animal Cognition* 18, no. 5 (2015): 1039–47. http://link.springer.com/article/10.1007%2Fs10071-015-0872-2.

Sentience Institute. "Momentum vs. Complacency from Welfare Reforms." In *Summary of Evidence for Foundational Questions in Effective Animal Advocacy.* https://www.sentienceinstitute.org/foundational-questions-summaries#momentum-vs.-complacency-from-welfare-reforms. Accessed November 9, 2017.

———. "Social Change vs. Food Technology." In *Summary of Evidence for Foundational Questions in Effective Animal Advocacy.* http://www.sentienceinstitute.org/foundational-questions-summaries#social-change-vs.-food-technology. Accessed November 9, 2017.

Serpell, James A. "How Social Trends Influence Pet Ownership (& Vice Versa)." Slideshow presentation. Website of the European Commission. http://ec.europa.eu/.

Sethu, Harish. "How Many Animals Does a Vegetarian Save?" Counting Animals, February 6, 2012. http://www.countinganimals.com/how-many-animals-does-a-vegetarian-save/.

———. "Meat Consumption Patterns by Race and Gender." Counting Animals, August 23, 2012. http://www.countinganimals.com/meat-consumption-patterns-by-race-and-gender/.

Shurtleff, William, and Akiko Aoyagi. "History of Tempeh and Tempeh Products (1815–2011)." Soyinfo Center, October 9, 2011. http://www.soyinfocenter.com/books/148.

———. "History of Tofu." Soyinfo Center. Accessed November 9, 2017. http://www.soyinfocenter.com/HSS/tofu1.php.

———. "History of Yuba—The Film That Forms atop Heated Soymilk (1587–2012)." Soyinfo Center, November 1, 2012. http://www.soyinfocenter.com/books/159.

———. "Madison College and Madison Foods: Work with Soy." Soyinfo Center. 2004. http://www.soyinfocenter.com/HSS/madison_college_and_foods.php.

Siegrist, Michael, and Bernadette Sütterlin. "Importance of Perceived Naturalness for Acceptance of Food Additives and Cultured Meat." *Appetite* 113 (2017): 320–26. https://www.ncbi.nlm.nih.gov/pubmed/28315418.

Smith, Andrew F. "Factory Farms." In *The Oxford Encyclopedia of Food and Drink in America*, vol. 2, 2nd ed. Edited by Andrew F. Smith. Oxford, UK: Oxford University Press, 2013.

Snow, Mary Lydia Hastings Arnold. *Mechanical Vibration: Its Physiological Application in Therapeutics*. New York: Scientific Authors' Publishing Co., 1912.

Specht, Liz, and Christie Lagally. "Mapping Emerging Industries: Opportunities in Clean Meat." Good Food Institute, June 6, 2017. http://www.gfi.org/images/uploads/2017/06/Mapping-Emerging-Industries.pdf.

Spencer, Colin. *The Heretic's Feast: A History of Vegetarianism*. London: Fourth Estate Limited, 1993.

Spurio, Maurizio. *Particles and Astrophysics: A Multi-Messenger Approach*. Cham, Switzerland: Springer International, 2015.

Tandon, Ajay, et al. "Measuring Overall Health System Performance for 191 Countries." World Health Organization, GPE Discussion Paper Series 30 (2000). http://www.who.int/healthinfo/paper30.pdf.

Tang, Anne Lise, et al. "Calcium Absorption in Australian Osteopenic Post-Menopausal Women: An Acute Comparative Study of Fortified Soymilk to Cows' Milk." *Asia Pacific Journal of Clinical Nutrition* 19, no. 2 (2010): 243–49. https://www.ncbi.nlm.nih.gov/pubmed?term=20460239.

Tobey, Kristen. *Plowshares: Protest, Performance, and Religious Identity in the Nuclear Age*. University Park, PA: Pennsylvania State University Press, 2016.

Tomasik, Brian. "How Many Wild Animals Are There?" Essays on Reducing Suffering. Last modified October 26, 2017. http://reducing-suffering.org/how-many-wild-animals-are-there/.

———. "How Much Direct Suffering Is Caused by Various Animal Foods?" Essays on Reducing Suffering. Last modified August 25, 2017. http://reducing-suffering.org/how-much-direct-suffering-is-caused-by-various-animal-foods/.

Tracey, W. Daniel Jr., et al. "Painless, a Drosophila Gene Essential for Nociception." *Cell* 113, no. 2 (2003): 261–73. https://www.ncbi.nlm.nih.gov/pubmed/12705873.

Tuso, Philip J., Mohamed H. Ismail, Benjamin P. Ha, and Carole Bartolotto. "Nutritional Update for Physicians: Plant-Based Diets." *Permanente Journal* 17, no. 2 (2013): 61–66. http://www.thepermanentejournal.org/files/Spring2013/Nutrition.pdf.

Van Mensvoort, Koert, and Hendrik-Jan Grievink. *The In Vitro Meat Cookbook*. Amsterdam: BIS Publishers, 2014.

Wakslak, Cheryl J., et al. "Moral Outrage Mediates the Dampening Effect of System Justification on Support for Redistributive Social Policies."

Psychological Science 18, no. 3 (2007): 267–74. http://www.ncbi.nlm.nih.gov/pubmed/17444925.

Webb, Barbara. "Cognition in Insects." *Philosophical Transactions of the Royal Society B: Biological Sciences* 367, no. 1603 (2012): 2715–22. https://www.ncbi.nlm.nih.gov/pmc/articles/PMC3427554/.

Wild, Florian, et al. "The Evolution of a Plant-Based Alternative to Meat." *Agro Food Industry Hi Tech* 25, no. 1 (2014). https://www.wur.nl/upload_mm/e/3/c/72826cba-6b00–4066-b3dd-7ca1546e92e4_LikeMeat%20Wild,%20Czerny,%20Janssen%20et%20al%20Agro Food%20Ind%202014.pdf.

Williams, William Carlos. *Delmarva's Chicken Industry: 75 Years of Progress*. Georgetown, DE: Delmarva Poultry Industry, 1998.

Wirtz, John G., Johnny V. Sparks, and Thais M. Zimbres. "The Effect of Exposure to Sexual Appeals in Advertisements on Memory, Attitude, and Purchase Intention: A Meta-Analytic Review." *International Journal of Advertising* (2017): 1–31. http://www.tandfonline.com/doi/full/10.1080/02650487.2017.1334996.

Witwicki, Kelly. "Sentience Institute Global Farmed & Factory Farmed Animals Estimates." Sentience Institute. https://docs.google.com/spreadsheets/d/1Njl_GS7jDOELjOtywvk3thIFpW_v10uZ5APJl1KgaYo/. Accessed November 9, 2017.

———. "Social Movement Lessons from the British Antislavery Movement: With a Focus on Applications to the Movement Against Animal Farming," Sentience Institute, November 27, 2017. www.sentienceinstitute.org/lessons-from-the-british-antislavery-movement.

World Animal Protection. *Animal Protection Index*. http://api.worldanimalprotection.org/. Accessed November 12, 2017.

You, Xiaolin, et al. "A Survey of Chinese Citizens' Perceptions on Farm Animal Welfare." *PLOS ONE* 9, no. 10 (2014). https://www.ncbi.nlm.nih.gov/pmc/articles/PMC4196765/.

INDEX

ACTAsia, 137
activism. *See* advocacy strategies and movement; social change
advertising strategies: of animal agriculture industry, 22–23, 106; of food tech companies, 68–71, 101; of MFA, 96, 124; of PETA, 25, 37, 123, 172n23; of specialty farms, 107
advocacy strategies and movement: bold language use in, 69, 86, 110–11, 128; against chicken and egg industry, 32–33; CIWF's work, 31–32, 140–41; on climate change, 125, 132–33; confrontational strategy, 126–29; effects of successful, 33–35; in global animal-free movement, 135–44; *how vs. why* messaging, 123–25; HSUS's work, 26–27, 38–40, 123; on "humane" animal farming, xv–xvi, 103–4, 111–12; identifiable victim effect, 125–26; inclusiveness in, 131–35; individual *vs.* institutional messaging, 36, 113–21; MFA's work, 26, 27, 30–31, 57, 96; moral outrage, 118–19, 127–28; PETA's work, 24–25, 37, 81, 84, 123, 172n23; race and ethnicity in, 129–31; on social justifications, 95–103; in Taiwan, 137; trigger events for, 121–23; wearing animal costumes, 25, 128, 172n22. *See also* effective altruism; social change
"ag-gag" laws, 27–30
algae-based foods, xii, 62
almond milk, 66, 68, 70
Alphabet, xii. *See also* Google
altruism. *See* effective altruism
Al-Waleed Bin Talal bin Abdulaziz al Saud, Prince of Saudi Arabia, 142
American Civil Liberties Union, 28
American Egg Board (AEB), 22–23, 42–43
Amir, Hovav, 86
ancestral diet, 97–99, 184n10. *See also* food traditions
Anchel, David, 81
The Animal Activist's Handbook (Ball and Friedrich), 58
animal agriculture industry: advertising campaigns of, 22–23; "ag-gag" laws, 27–30; regulation of, 21–22; statistics on, 36–37; as term, 167n1. *See also* cellular agriculture industry; factory farming
Animal Charity Evaluators (ACE), 30–31, 85
Animal Equality (organization), 35–36, 142
animal farming: "humane" justification of, xv–xvi, 103–9, 111–12, 186n30; Neanderthal thought experiment on, 104–5; statistics on, ix–x, 36–37; as term, 167n1. *See also* animal agriculture industry; factory farming; *specific animal names*
animal-free fashion show, 137
animal-free food system: animal welfare as motivation for, xi, 32–35,

94, 169n1, 191n16; consumer research experiments on, 90–91, 101–2, 115; corporate support of, 54–56, 59–60; economic argument for, 34, 131, 134; future of, 157–64, 191n20; global conditions and movement, 135–44; Hampton Creek case study, 38–45; human health as motivation, xi, 21, 53, 62, 80, 84, 99, 156; impact of artisan companies on, 61–64; labeling standards for, 102–3; market selling of, 64–71; product evaluations of, 60–61, 109–11; social justifications against, 95–103; terminology in, 45, 67–68, 74, 88–91, 168n13. *See also* advocacy strategies and movement; cultured meat; morality

Animal Liberation (Singer), 114

animals: humans as, 167n1; legal protection for, 4, 11–12, 17–18, 23, 24, 156; religious beliefs about, 14–16; sanctuaries for suffering, 1–3, 17, 20, 161, 162; sentience of, x, xv, 3–4, 6–8, 61, 149, 152–53; as term, 167n1; who eat other animals, 32, 168n15. See also *specific animal names*

animal sanctuaries, 1–3, 17, 20, 161, 162

animal welfare: in China, 141; conditions in factory farming, x, 4–5; at "humane" farms, xv–xvi, 103–9, 111–12; in India, 139; as motivation for animal-free food system, xi, 32–35, 94, 169n1, 191n16; of wild animals, 149–52, 191n8

antibiotics, 108, 138, 167n1

anticonfinement legislation, 30–32, 34

antislavery movement, 33–34, 110, 116, 127, 186n7

antismoking campaigns, 22, 67–68, 102, 110, 116–17, 159

Aristotle, 3–4

artificial breeding, x, 98, 99, 100, 109, 138

artificial intelligence (AI), 92, 148, 157

artificial sentience, 154–55

Asch, Solomon, 96

athletes, xi, 97, 133–34

Australia, 88, 156

Austria, 160

Baboumian, Patrik, xi

Baby Jack's BBQ, 84

backfire effect, 127–28

Balk, Josh, 38–40, 86

Ball, Matt, 58

bananas, 98

battery-cage chicken farms, 17–20, 114, 139. *See also* chicken farming

BBC, ix

beagles, 2

bean proteins, 41. *See also* soy products

bearbaiting, 4

Beatles (band), 14–15

beef. *See* cow farming; cultured meat; meat alternatives

beer, 89

Benjaminson, Morris, 74

Berns, Gregory, 10

Besançon, Michael, 61

"The Best Speech You'll Ever Hear" (video by Yourofsky), 86

The Better Angels of Our Nature (Pinker), 13

Beyond Meat, 50–51, 54, 67–68, 71, 141

Bin Alwaleed bin Talal, Prince Khaled, of Saudi Arabia, 142–43

BioCurious, 82

biomedical research, 82. *See also* scientific research on animals

bioreactors, 92

Bittman, Mark, 28

bivalves, 61

Blackfish (film), 10–11

Black Lives Matter, 122

Blakely, Sara, 44

Bloomberg, 43

bold language strategy, 69, 86, 110–11, 128
Bollard, Lewis, 136, 138, 139, 141
Bolt Threads, 82
Boot, Johan, 41
Bouazizi, Mohamed, 122
Boyman, Todd, 51–52
Branson, Richard, ix
Brazil, 58
bread, 2, 65
breeding. *See* artificial breeding
Brin, Sergey, xii, 76
Brown, Ethan, 50–51, 54, 67
Brown, Patrick "Pat," 49–50, 54, 70–71
Brown v. Board of Education, 133
Buddhism, 14–15, 137, 141
bugs, 152–53, 167n1, 167n3
Burgess, Allison, 52
butter alternatives, 65
Buttercup (cultured yeast), 79
buyback programs, 43–44

cage-free policies, 30–32, 34
California Proposition 2, 17–18, 27, 34
Cambridge Declaration on Consciousness (2012), 9
Canada, 140
cannibalism, 32, 168n15
Cargill, 90
Carnage (film), ix
Carrel, Alexis, 73
Carter, David, xi
casein, 79
cashew milk, 70
cats, 4, 12–13
cattle. *See* cow farming
Catts, Oron, 74–75
"cell-cultured meat" as product term, 74, 91. *See also* cultured meat
cell culture media technology, 92
cell lines, 93
cellular agriculture industry, xv, 70–71, 73–75, 81–82, 91–92. *See also* animal agriculture industry; cultured meat; *specific product types*
Chang, David, 50
Change.org, 4
Chatfield, Tom, ix
"check off" programs, 22–23
cheese alternatives: from Follow Your Heart, 62; from Kite Hill, 50, 54; from Miyoko's Kitchen, 63–64, 64–65; from Not Company, 142; Real Vegan Cheese, 82. *See also* dairy products
chicken alternatives: Beyond Meat chicken strips, 51; of MAF, 87; Memphis Meats, 83; R&D in, 81, 85–86
chicken farming: abusive conditions in, 4–5; battery-cage farms, 17–20, 114, 139; cage-free policies, 30–32, 34; in China, 137–38; diseases, 106; "humane" or specialty farms, 105–9; in India, 139; investigations of, 17–20, 25–26, 38–39, 105–9; production evaluation of, 60, 98, 178n4; statistics on, ix, 137. *See also* factory farming
Chile, 142
chimpanzees, 11–12
China: animal-product consumption in, 58, 136–37, 143; animal welfare and movement in, 137–38, 141; environmental policies in, 140; infant formula scandal from, 65–66, 141; Tiananmen Square protests, 122, 137; tofu consumption in, 46; vegetarianism in, 141–42
Christianity, 15
Churchill, Winston, 73
Cialdini, Robert, 112
cigarette industry. *See* antismoking campaigns
civil disobedience, 122, 126–27, 137. *See also* collective action; individual *vs.* institutional change
Clara Foods, 81
clean energy industry, 89, 91, 159
"clean meat" as product term, 74, 89–91. *See also* cultured meat
Clem, Will, 84

climate change advocacy, 125, 132–33. *See also* environmental movement
clothing industry, 82, 122. *See also* leather
CNBC, 43
CNN, 29
coconut milk, 68, 70
collagen, 81–82
"collapse of compassion" theory, 117–18, 126
collective action, 113–15, 117–18. *See also* civil disobedience; individual *vs.* institutional change
"collective effervescence," 117
Coller, Jeremy, 55
Coman-Hidy, David, 30
Commonwealth Enterprises, 24
communism, 132, 136
Compassion in World Farming (CIWF), 31–32, 140–41, 186n10
Compassion Over Killing (COK), 25
compassion theory, 117–18
confrontational activism, 126–28
conservative *vs.* liberal political arguments, 131–33
consumer behavior: complacency, 33; individual *vs.* institutional change in, 61, 113–21; justification of the *how vs. why*, 123–24; "natural" justification for, 97–102; "necessary" justification for, 97; "nice" justification for, 97; prices and effects on, 34–35, 131, 134; psychological refuge, 109–11; research on cultured meat terminology for, 90–91, 102; risk acceptability survey, 101–2; social pressure, 56, 95–96, 113–15, 119–20
Cooney, Nick, 58
Coons, Derek, 58
Copernicus, Nicolaus, 5
corn, 98
corporate investments: in cultured meat, 85–86, 90; in cultured milk, 78–79; in plant-based products, 54–56, 59–60; for social change, xii, 40–41, 94
corporate responsibility, advocacy for, 31–33, 39–40, 55–56
costumes used in animal rights advocacy, 25, 128, 172n22
Couette cell, 52–53
Counter Culture Labs, 82
cow farming: pasture-raised *vs.* factory farmed, 105–9; production evaluation of, 60; slaughterhouse investigations of, 27, 28. *See also* factory farming
coyote, 15
cultured eggs, 81
cultured meat, xi, xii; Catts's sheep exhibit, 74–75; early messaging of, 101; future of, 92–94; by Hampton Creek, 84–87; by Israeli companies, 86–87; by MAF, 87; Memphis Meats, xi, 59, 83–85, 90; MosaMeat, 83; NASA's goldfish experiment, 74, 77; PETA's contest for, 81; production history of, 73–74; production transparency of, 102; public taste testing of, 73, 76–77, 123; Shojinmeat Project on, 101; terminology for, 45, 67–68, 74, 88–91, 102; by Tyson Foods, 85–86. *See also* cellular agriculture industry; meat alternatives
cultured milk, 78–81, 89

The Daily Show, 29
dairy alternatives: cultured butter, 65; cultured milk, 78–81, 89; eggless mayonnaise, xii, 41–45, 54–55, 62; plant based cheese, 50, 62–65, 142; plant-based milk, 66–67; plant-based yogurt, 54, 80–81, 89
dairy products: labeling of, 64–66; production evaluation of, 60; in Taiwan, 137. *See also* milk products
Darwin, Charles, 6, 8
Datar, Isha, 78, 81
Dean Foods, 71
Democracy Now!, 28

Descartes, René, 3, 4
"descriptive norm," 120
de Waal, Frans, 9
Diaz, Nate, xi
digital sentience, 154–55
Diners, Drive-Ins, and Dives (TV show), 72
disease management, 20, 108, 138, 167n1
Disney, Walt, 44
Dodds, Peter, 95
dogs, 2, 3, 12–13, 16, 143
"Dogs Are People, Too" (Berns), 10
dominion, as term, 15
duck alternatives, 83
duck farming, 24
Durkheim, Émile, 117

Earth Island, 62
Earthrise, 158
earthworms, 152–53
Eating Animals (Foer), 58, 79
economic argument for animal-free foods, 34–35, 131, 134
effective altruism: authors experience with, xiii–xiv, 27; charities who practice, 57, 78, 85, 135–36; global conference on, 83; prioritization framework in, 148–49. See also *specific advocacy organizations*
egg alternatives: cultured eggs, 81; eggless mayonnaise, xii, 41–45, 54–55, 62, 142; meringue, 81; VeganEgg, 62
eggs, 60, 138, 178n4. *See also* chicken farming
elephants, 6
Elizondo, Arturo, 81
energy industry, 89, 91, 122, 159
environmental movement, 95, 116, 122, 125, 140. *See also* climate change advocacy
Environmental Performance Index, 140
Erlich, Daniel, 86
ethnicity and animal-free movement, 129–30
The Expression of the Emotions in Man and Animals (Darwin), 6
Exxon *Valdez* oil spill (1989), 122

factory farming: artificial breeding in, x, 98, 99, 100, 109, 138; battery-cage chicken farms, 17–20, 114, 139; in China, 137–38; conditions of, x, 4–5; governmental subsidies for, 21–22; in India, 77, 136, 138–39; investigations of, 23–24; legislative bills on, 17–18, 23; statistics on, ix–x, 17, 109; synthetic chemical use in, 98–99; as term, 167n1. *See also* animal agriculture industry; slaughterhouse investigations
Farm Animal Investment Risk & Return (FAIRR), 55
farm sanctuaries, 1–3, 17, 20, 161, 162
Farm Sanctuary, 58
fashion industry, 137. *See also* garment industry; leather
Fast Action Network, 164
Faunalytics, 130
FDA (Federal Drug Administration), 42
Feedstuffs, 33
feminization and moral concern for animals, 13, 146
fermented products, 89
Fetherstonhaugh, Rob, 75–76
Fifty Years, 56
"Fifty Years Hence" (Churchill), 73
Finless Foods, 87
Fischer, Eitan, 85
fish alternatives, xii, 87–88
fish farming: conditions in, 4, 185n26; fish consumption statistics, ix; production evaluation of, 60, 110, 178n4
Five Precepts of Buddhism, 14–15
Foer, Jonathan Safran, 79
foie gras farm investigation, 24
Follow Your Heart (store and brand), 61–63

food advocates. *See* advocacy strategies and movement
Food Frontier, 88
food traditions, 97–99. *See also* ancestral diet
France, 122, 140
Friedrich, Bruce, 58–59
frog meat, 75
fruit flies, 152
Future Meat Technologies, 87
future speculation, 147–49, 154–55, 157–64, 191n20

Gandhi, Maneka, ix
Gandhi, Mohandas (Mahatma), 14, 122
Gandhi, Perumal, 78–79, 80
Garces, Leah, 31–32
garment industry, 82, 122, 137. *See also* leather
Garrison, William Lloyd, 116
gelatin, 81
Geltor, 81
gender and animal-free food movement, 129–30, 134
General Mills, 40, 54
genetically modified organism (GMO), 53, 100–101
Genovese, Nicholas, 84
Germany, 140
Gilboa, Tal, 86
GiveWell, 135–36
Global Animal Partnership 5-Step program, 55
globalization and moral concern for animals, 13, 135–44
goat milk, 66
Goldberg, Bob, 61–62
goldfish cultured meat experiment, 74, 77
Good Food Institute (GFI): establishment of, 57–60; In-N-Out Burger campaign by, 119; research focus and funding by, 57–58, 77, 92; on terminology, 89, 91. *See also* Mercy for Animals (MFA)
Good Ventures, 136
Google, xii, 76, 83
goose farm, 24
Graham, Jesse, 132
Graham, Sylvester, xi
Grande, Ariana, 96
Grandin, Temple, 7, 141
grass-fed *vs.* grain-fed cow farming, 107–8
Great Britain: abolition of slavery in, 33–34, 110, 116, 127, 161, 186n7; mockumentary on eating animals in, ix
Greek philosophy, 3–4
Green Is the New Red (Potter), 28
Green Monday (company), 141
Gregory, Chad, 32–33
grief behavior, 7
Griffin, Donald, 9
grocery stores, 43, 50–51, 55, 71. *See also* labeling standards; Whole Foods
Guardian, 42–43, 72
gun regulation, 21–22

Haidt, Jonathan, 132
hamburgers. *See* cow farming; cultured meat; meat alternatives
Hampton Creek, 38–45, 84–86, 87, 123, 142
Hanyu, Yuki, 101
Harari, Yuval Noah, x
Harvest Home Animal Sanctuary, 1–3
heme, 49–50, 51, 79
herbicide use, 98–99
Herbivorous Butcher, 71–72
Hercules (chimpanzee), 11–12
Herzog, Hal, 9
Hinduism, 14, 15–16, 139
History of Meat Alternatives (965 CE to 2014) (Shurtleff), 45–46
Hog Farm Management, 21
Holland, Freddie, 52
Homo Deus (Harari), x
honeybees, 152–53
Hong Kong, 141
Horizons Ventures, 78

household cleaning products, 44, 69
Huli Huli Hawaiian ribs, 72
"humane" animal farming, justification of, xv–xvi, 103–9, 111–12, 186n30
Humane League, 30, 143, 164
Humane Society of the United States (HSUS), 26–27, 123; Balk's work for, 38, 39–40
human health as motivation for animal-free diet, xi, 21, 53, 62, 80, 84, 99, 156, 159
human population, future, 147, 154–55, 191n18, 191n20
humans as animals, 167n1
hunger strike, 122
Hungry Planet, 51–52, 158

Idaho, 28
identifiable victim effect, 125–26
Impossible Burger, 47–48
Impossible Foods, xii, 47–48, 49–50
inclusive advocacy strategies, 131–35
India: animal farming in, 77, 136, 138–39; independence of, 14; meat consumption in, 52, 58, 83; Salt March, 122; vegetarianism in, 14, 36, 80, 138
IndieBio, 81
individual *vs.* institutional change, 36, 113–21
Indonesia, 46
industrial farming. *See* factory farming
infant formula, 65, 141
Influence: The Psychology of Persuasion (Cialdini), 112
"injunctive norm," 120
In-N-Out Burger, 119
Innovative Food Science and Emerging Technologies, 78
insects, 152–53, 167n1, 167n3
institutional *vs.* individual change, 36, 113–21
Integriculture, 101
interstellar colonization, 147
"in vitro meat" as product term, 74, 75, 88. *See also* cultured meat
The In Vitro Meat Cookbook, 75
Iowa, 28
Israeli activists, 86–87, 96

Jainism, 14
Japan, 46, 101, 140
Jhala, Ravi, 80–81
Jones, Chris, 41
Juarez, Bernadette, 108
The Jungle (Sinclair), 24
JUST. *See* Hampton Creek
Just Mayo, xii, 41, 42–45

Kaiser Permanente, 97
Kellogg, John Harvey, xi
Kickstarter campaigns, 72
killer whales, 8–9, 10–11
Kite Hill, 50, 54
Klein, Ezra, ix
Klein, Joshua, 41
Ko, Aph, 130

labeling standards: for "all foods containing DNA," 100; for animal-free products, 102–3; for cigarettes, 67–68, 102, 159; for dairy products, 64–68; for mayonnaise, 45; for meat, 67–68. *See also* grocery stores
"lab-grown meat" as product term, 45, 74, 88–89. *See also* cultured meat
LAL (limulus amebocyte lysate), 82
leather, 60, 61, 82
Leo (chimpanzee), 11–12
Lewin, Paul, 61
liberal *vs.* conservative political arguments, 131–33
Li Ka-shing, 38, 78
London Vegetarian Society, 14
Lorestani, Alex, 81–82
Los Angeles Times, 66
Lowry, Adam, 44, 69
lupin, 53
Lusk, Jayson, 107–8

Mackey, John, 55
Marek's disease, 106
marketing strategies. *See* advertising strategies; advocacy strategies and movement
Massachusetts, 23
Masson, Jeffrey, 9
Matheny, Jason, 77, 78, 83, 182n25
mayonnaise, xii, 41–45, 54–55, 62, 142
McClure, Jessica, 125–26
McCruelty campaign (PETA), 24–25
McDonald's, 24, 114, 141
meat alternatives: Beyond Burgers, 50–51, 67–68, 141; Beyond Meat chicken strips, 51; Couette cell, 52–53; grocery-store brands of, 55; heme in, 49–50, 51, 79; of Herbivorous Butcher, 72; history of, 45–46, 47–54; Hungry Planet burgers, 51–52; Impossible Burgers, 47–48; "tofurkey," xv, 46–47; veggie burgers, xv, 46. *See also* cultured meat
meat boycott (US), 20
meat recall (US), 123
Meat the Future, 87
"meat," use of term in products, 67–68, 88
media: effects on ethical industry companies, 45, 47; on GFI, 59; on Hampton Creek practices, 42–44; on Herbivorous Butcher, 72
medical research. *See* scientific research on animals
medicine used on animals, 20, 108, 167n1
Memphis Meats, xi, 60, 83–85, 90
Mercy For Animals (MFA), 26, 27, 30–31, 57, 96. *See also* Good Food Institute (GFI)
Meredith, Emily, 28–29
meringue, 81
methane, 108
The Method Method (Lowry), 44
Middle East, 142
Milgram, Stanley, 96
milk alternatives: cultured milk, 78–81, 89; plant-based, 66–67; soy milk, 46, 66, 68, 71
milk products: labeling standards for, 65–66, 67, 68; pasteurization and, 21. *See also* dairy products
Missouri, 22
Miyoko's Kitchen, 63–64, 142
Modern Agriculture Foundation (MAF), 87
Modern Meadow, 82
morality: far future speculation and, 147–49, 154–55, 157, 191n20; moral foundations theory, 132; moral outrage, 118–19, 127–28; moral progress overview, 16, 145–46; Neanderthal thought experiment, 104–5; religious beliefs about animals, 14–16
MosaMeat, 83
Moskovitz, Dustin, 136
Mrs. Wilmer Steele's Broiler House, 20
mung bean protein, 41
Music Lab experiment, 95–96
Musk, Elon, 147
mussels, 61
Muufri. *See* Perfect Day (company)
Mwalua, Patrick Kilonzo, 150, 190n7

NASA, 59, 74, 77, 155
National Pork Producers Council, 30
National Press Photographers Association, 28
National Review, 21
Native American stories about animals, 15
"natural" justification for animal-based diet, 97–102
Neanderthal thought experiment, 104–5
"necessary" justification for animal-based diet, 97
Netherlands, 35, 75, 77, 181n12

New Crop Capital (NCC), 60
New Harvest, 78–81, 83, 89, 91–92
New Wave Foods, xii
New York Times, 10, 28, 44, 107–8
"nice" justification for animal-based diet, 97
Nonhuman Rights Project (NhRP), 11–12, 156
nonviolence and animal cruelty, 14
"normal" justification for animal-based diet, 95–96
North Africa, 142
NotCheese, 142
Not Company, 142
NotMayo, 142
NPR, 72
nuclear energy movement, 122
nut products: almond milk, 66, 68, 70; cashew milk, 70; cheese, 50, 54, 63–64; coconut milk, 68, 70
Nye, Bill, ix

octopuses, 7–8
oil industry, 122
Okamoto, Toni, 134
Okja (film), 123
olive oil scandal, 65
On the Revolutions of the Heavenly Spheres (Copernicus), 5
Open Philanthropy Project, 135–36, 141
Open Wing Alliance, 143
Orange Is the New Black (TV show), 130
orcas, 8–9, 10–11
oysters, 61

Pacelle, Wayne, 27
Pandya, Ryan, 78–80
papaya industry, 100
Papua New Guinea, 48–49
parsimony, 5–6, 8
pasteurization, 21
pasture-raised animals, 107–8
patriotic messaging, 132–33, 139
pea protein, 52–53, 69
Pei Su, 136
People for the Ethical Treatment of Animals (PETA), 24–25, 37, 58, 81, 84, 123, 172n23
Perdue (company), 31–32, 86
Perfect Day (company), 78–81
persuasion, 112
pesticide use, 53, 98–99, 122
pet ownership, 12–13
pig farming, 4, 21, 24, 34. *See also* factory farming
pigs, 2, 6–7
Pinker, Steven, ix, 13
plant-based, as term, 168n13
Plant Based Foods Association, 53
Plant Based on a Budget (Okamoto), 134
Plato, 3–4
A Plea for Vegetarianism (Salt), 14
Plowshares movement, 59
Post, Mark, 73, 75, 83, 181n12
Potter, Will, 28–29
poultry alternatives, xv, 46–47, 83, 99. *See also* chicken alternatives
poultry farming, 1, 24, 60, 107, 178n4. *See also* chicken farming
"problem of other minds" theory, 8
Proposition 2 (California), 2, 17–18, 27, 34
protein consumption, 68, 70, 180n24. See also *specific food products; specific protein types*

Quartz, 21
The Question of Animal Awareness (Griffin), 9

race and animal-free food movement, 129–31
racism, 122, 130, 133. *See also* slavery, abolition of
rats, 6, 16
Real Vegan Cheese, 82
reducetarians, 36–37
religious vegetarianism, 14–16, 36, 137, 139, 141
renewable fuel, 69
Renninger, Neil, 69

Richmond Times-Dispatch, 40
The Righteous Mind (Haidt and Graham), 132
"right to farm" legislation, 22–23
Ripple Foods, 44, 69–70
Roman civilization, 4, 65
Ross, Russell, 73–74
Roux, Wilhelm, 73
Runkle, Nathan, 26, 58
Russia, 140

Sacks, Oliver, 7
Safeway, 49
Salganik, Matthew, 95
Salt, Henry, 14
Salt March (India), 122
sanctuaries, farm, 1–3, 17, 20, 161, 162
Sapiens (Harari), x
Saudi Arabia, 142–43
sausage, plant-based, 72
scaffolding (cultured meat technology), 93
Schinner, Miyoko, 63–64, 64–65
Schmidt, Eric, xii
school cafeterias, 158, 159
Scientific American, 99
scientific research on animals, 2, 6–7, 11–12, 16. *See also* biomedical research
scientific revolution, 5–6
seafood alternatives, xii, 87–88. *See also* fish farming
SeaWorld, 10
Seed (restaurant), 46
seitan, 46
sentience, artificial, 154–55
Sentience Institute, 164
sentience of animals, x, xv, 3–4, 6–8, 60–61, 149, 152–53
Seventh-Day Adventists, 46
Shahi, Houman, 158
Shapiro, Paul, 21, 27
Shavit, Ori, 86
sheep meat, 74–75
shelf placement in grocery stores, 71
Shojinmeat Project, 101
Shooster, Jay, 110
shrimp alternatives, xii
Shurtleff, William, 45–46
Silent Spring (Carson), 122
Silk (milk brand), 71
silk products, 82
Simon, Michele, 53
Sinclair, Upton, 24
Singapore, 122
Singer, Peter, 114
skin testing on animals, 2
slaughterhouse investigations: by Animal Equality, 35–36, 142; of chickens, 17–20, 25–26, 38–39, 105–9; of cows, 27, 28; by HSUS, 123; of pigs, 24; in Taiwan, 137. *See also* factory farming
slavery, abolition of, 33–34, 110, 116, 127, 161, 186n7
Smithfield, 40
social change: challenges to animal-free diet, 95–103; corporate-led, xii, 40–41, 94; history of, x–xi, 4, 41–42, 116–17, 127; individual *vs.* institutional action for, 36, 113–21; *vs.* technological change, 183n41; from trigger events, 121–23. *See also* advocacy strategies and movement
social pressure, 56, 95–96, 113–15, 119–20
Sothic Bioscience, 82
South Korea, 52
soy products: labeling milk from, 66–67; soy milk, 46, 66, 68, 71; tempeh, 46; tofu, 45–46, 142; tofurkey, xv, 46–47
space colonies, 147
Spanx, 44
specialty farms, xv–xvi, 103–9, 111–12, 186n30
"spent" hens, 17, 18–19
Spiber, 82
spiders, 152–53
Steele, Cecile, 20

story-based approach, 125–26
Stray Dog Capital, 56
Sunday Times, 75
sunflower oil scandal, 65
SuperMeat, 69, 87
Switzerland, 160
synthetic beef hamburger. *See* cultured meat
synthetic chemical use, 98–99

tail-docking practices, 30
Taiwan, 136–37
Taylor, Thomas, 146
Technological Readiness Assessments, 59
technological *vs.* social change, 183n41
Telegraph, 72
tempeh, 46
Tesla, 64
Tetrick, Josh, 38, 40–45
Theodosian Code, 65
Tiananmen Square protest (China), 122, 137
Tibbot, Seth, 49
Tilikum (orca), 10–11
Time, 72
tobacco regulation, 22. *See also* anti-smoking campaigns
tofu, 45–46, 142
"tofurkey" (product), xv, 46–47
Tofurky (brand), 46–47, 49
tool use of animals, 7–8
Triangle Shirtwaist Factory fire (1911), 122
trigger events, 121–23
Tuna, Cari, 136
turkey alternatives, xv, 46–47, 99
turkey farming, 1, 60, 107, 178n4
Turtle Island Foods, 46–47
Tyson Foods, 54, 55–56, 85–86

Unilever, 42, 54–55
United Egg Producers, 32–33
United Nations Food and Agriculture Organization and Fish Count, 136
urbanization and pet ownership, 12–13
USDA (US Department of Agriculture), 108; dietary guidelines by, 21, 159
US Health and Human Services dietary guidelines, 21
US News & World Report, 139
US Securities and Exchange Commission, 43
Utah, 28

Valeti, Uma, 83–84
Van Eelen, Willem, 74, 75
VeganEgg, 62
veganism: of athletes, xi, 97, 133–34; bivalvegans, 61; ethical omnivores *vs.*, 110; historical acceptance of, xvi; of Israeli activists, 86; race and media coverage of, 130; statistics on, 115
"vegan mafia," 56
VegeBurger, 46
Vegenaise, 62
vegetarianism: for ascetic purposes, 129; of athletes, xi, 97, 133–34; correlation between animal welfare standards and, 35; of famous people, 14–15, 96; Gandhi on, 14; historical acceptance of, xi, xvi, 47; in Hong Kong, 141; in India, 36, 80, 138; as individual act, 36, 115, 117; MFA ads for, 124; religious-based, 14–16, 36, 137, 139, 141; statistics on, 36–37, 115, 129–30; stereotypes of, 129–31, 133–34
Vegetarian Resource Group, 36, 129–30
veggie burgers, xv, 46
Verstrate, Peter, 75, 83
A Vindication of the Rights of Brutes (Taylor), 146
A Vindication of the Rights of Women (Wollstonecraft), 146
virtual reality, 35–36
vitamin-enriched rice, 100

Walch, Aubry and Kale, 71–72
Washington Post, 25
Watts, Craig, 31–32
Watts, Duncan, 95
Wegmans, 55
Weissmuller, Johnny, 62
weltschmerz, 163
whales, 8–9, 10–11
When Elephants Weep (Masson), 9
whey, 79
White Wave, 71
Whitts Barbecue, 84
Whole Foods, 43, 50–51, 55, 71, 107
"Why Industrial Farms Are Good for the Environment" (Lusk), 107–8
wild animals, 149–52, 191n8
Williams, Serena, xi
Williams, Venus, xi
Windbiel, Spencer, 61
Wollstonecraft, Mary, 146
wolves, 15
World Animal Protection, 140
World Anti-Slavery Convention, 116
World War I, 20
worms, 8

Yahoo! Finance, 66
yeast, 50–51, 79, 101
yellow pea protein, 52, 69
"Yes on 3" campaign, 23
Yeung, David, 141
Yoerg, Sonja, 9
yogurt, 54, 80, 89
"You Can Save the Planet" (Tetrick), 40
Yourofsky, Gary, 86
yuba, 46
Yucatan minipig, 2

Zahn, Oliver, 47–48